The Graduate Student's

Backpack

It's What You Need on the Research Path

The Graduate Student's

Backpack

It's What You Need on the Research Path

Ronald D. Taskey

Illustrated by Tina Vander Hoek

Book and Multimedia Publishing Committee
   April Ulery, Chair
   Warren Dick, ASA Editor-in-Chief
   E. Charles Brummer, CSSA Editor-in-Chief
   Andrew Sharpley, SSSA Editor-in-Chief
   Mary Savin, ASA Representative
   Mike Casler, CSSA Representative
   David Clay, SSSA Representative

Lisa Al-Amoodi, Managing Editor
Gail Schmitt, Editor

American Society
of Agronomy

Soil
Science
Society of America

Crop Science
Society of America

American Society of Agronomy
Soil Science Society of America
Crop Science Society of America, Inc.
5585 Guilford Road, Madison, WI 53711-5801 USA

agronomy.org/publications/books
soils.org/publications/books
crops.org/publications/books
www.SocietyStore.org

ISBN: 978-0-89118-334-1
e-ISBN: 978-0-89118-335-8
doi:10.2134/2012.graduate-students-backpack
Library of Congress Control Number: 2012935129

ACSESS publications
ISSN 2165-9834 (print)
ISSN 2165-9842 (online)

Cover design: Patricia Scullion
Interior design: Kristyn Kalnes
Cover and inside illustrations by Tina Vander Hoek

CONTENTS

FOREWORD

Clearly, the choice to pursue graduate studies is a decision that can have tremendous ramifications on one's professional development. The journey from structured undergraduate degree programming to largely unstructured graduate education is a significant and deliberate paradigm shift, one that few students are well equipped to manage. This book provides a practical guide to the graduate education experience, beginning with the critical task of choosing graduate advisers, committee members, and a research topic, and ending with how to present yourself to employers when you have completed your graduate studies. Every student contemplating the pursuit of an M.S. or Ph.D. degree should use this book as a guide.

Dr. Taskey has provided very useful universal hints on becoming a successful graduate student and "closing the deal" with your thesis and scholarly publications. This book examines the personal and professional challenges that most graduate students encounter and offers insightful solutions to these trials and tribulations. It also provides direction on developing resumes and interviewing skills.

We believe that graduate students and graduate-student advisers will frequently refer to this publication. Before long, this guide to graduate school will feel as comfortable as that backpack you carry every day. Enjoy the journey through this book.

Ken Barbarick, 2012 American Society of Agronomy President
Gary Pierzynski, 2012 Soil Science Society of America President
Jeffrey J. Volenec, 2012 Crop Science Society of America President

PREFACE

New graduate students usually begin their advanced academic career in a state of excitement and anticipation. They have before them an opportunity that only a tiny fraction of the world's people can experience—indeed, they are privileged individuals. As they settle in, meet new people, and learn their new environment, they begin to grasp the reality of their situation. They begin to see the responsibilities that go with privilege, and all too soon the process of earning a research-oriented advanced degree looms as arduous and confusing.

The new student must select a graduate committee, plan course work, find an original research topic, and design a research plan that leads to a defensible thesis or series of publishable research articles. Whether the student plans a research career or not, he or she typically must complete an original, independent, analytical study, even though few beginning students know how or where to begin. Clearly, the student needs help.

This process-oriented manual holds the fundamentals students in the sciences need to trek the graduate school path—it is their "backpack." Here's what it will do for you:

- Suggest ways to get started with a graduate committee, select a research topic, and deal with the challenges of life as a graduate student.

- Outline the requirements of quality graduate research, summarize research philosophies, and present intellectual tools to inspire and exploit your creativity.

- Present a time- and student-tested model for research planning that will put you firmly on the path to success.

- Spell out techniques for finding, reading, and evaluating scientific reports and offer reliable hints to help you write clearly and concisely.

- Show you how to keep track as you struggle to keep on track.

- Provide pointers to help you favorably present yourself and your work to your graduate committee and the broader scientific community.

- Offer a final thought on the hereafter of graduate school—the so-called but misnamed "real world." (When you're in it, graduate school is the real world. Don't let anyone convince you it's not.)

What it will not do: This is not a research methods book. It offers no instruction in experimental protocols, designs, or analyses. Those are related but distinctly different subjects that can vary greatly among disciplines and about which numerous excellent volumes have been written. Rather, this handbook is intended to come before the methods manuals. It addresses needs, concerns, and concepts that are common to nearly all graduate students.

Use it, and this guide will serve you well—whether your interests are in physical science, life science, social science, or engineering and whether you pursue a master's degree or a doctorate. For the most part, the problems and challenges dealt with are universal. Still, you will scale your own obstacles and rejoice in your own triumphs, and the timing and intensity of downs and ups and deadlines will vary among you and your colleagues. So go forth and back through the chapters, scribbling notes in the margins to suit your needs. The more worn and dog-eared your book becomes, the better.

Numerous people spurred me on and offered helpful critique in developing this book. First, thanks go to hundreds of former students who encouraged—and to a few who discouraged—me in this endeavor. While the encouragements kept me going, the discouragements led to a better product. I'm grateful to both groups and I wish them well.

A great deal of credit and my heartfelt thanks go to mentors from my own graduate-student days. Although there were many, on the topics in this book I feel most indebted to Dr. Thomas Nimlos, Dr. Robert Wambach, and Dr. Moyle Harward. Numerous friends and colleagues—at least one of whom repeatedly discussed some of the philosophical points well into the night—offered valuable comments and encouragement: Dr. Merton Richards, Dr. J. Michael Kelly, Dr. Robert Graham, Dr. Thomas Rice, Dr. William Preston, Dr. Lynn Moody, and Dr. George Cotkin. Finally, the artistry of Ms. Tina Vander Hoek injected levity into some otherwise humorless topics. It's good to have good people on your side.

And now that it's done, I feel ready to begin.

Ronald D. Taskey

CHAPTER 1

The Beginning

What's This Chapter About?

What's graduate school all about?

How do I find advisors, and what do they do?

What about a research topic?

What other things should I think about?

What about exams?

What's In This Chapter?

First, You Need Advisors—and They Need Students

Selecting an Advisor
Inducing an Advisor to Select You
Selecting a Graduate Committee
Getting the Most from Your Committee
Your Responsibilities to Your Advisor, Your Committee, and
Your Group

Next, You Need a Research Topic and a Strategy

Criteria for Identifying a Suitable Topic
Recognizing and Selecting a Researchable Goal or Problem
Setting Working Goals and Leaving Tracks

Some Personal and Practical Points

On Being a Graduate Student
Communicating
Academic and Professional Goals and Objectives
Time Management
Money
Ups and Downs
Stress-Strain Relationships
Suffer It
Prevent or Decrease It
Deflect or Dissipate It
Strengthen Your Resistance
Remove Yourself from Its Path
Advice on the Mundane Side

Major Hurdles: Preliminary Exams and Final Defense

Preliminary or Qualifying Examinations

What Are They?
What to Expect
How to Prepare
Possible Outcomes and Consequences

The Final Defense

"To see a World in a Grain of Sand..."

William Blake, *Auguries of Innocence*

When a graduate student—faced with the challenges of formulating, conducting, writing, and defending a thesis or professional paper— asks the question, Where do I begin? chances are she or he already has begun. If you are that student, you at least have chosen a field of study, and because you are reading this, you have started the thought process. If you have committed yourself to a scientific research-oriented degree, you have begun a quest to see beyond what you have seen before, and perhaps beyond what others have seen. But to be truly successful, you must do more than see beyond: you must go there.

Getting there is the hard part, as it should be. Getting there can follow a tortuous path, fraught with trials, pitfalls, barriers, and frequent long days and nights of work. Sometimes the path runs a sluggish, confusing course seemingly designed to do little more than induce stress and test your endurance. Other times it runs smooth and satisfying, and you delight in the trek. Whether through turbulence or tranquility, the path will cross nuggets for you to discover along the way. Unearth enough nuggets, learn to remold them into a defensible thesis or professional paper, and you can emerge with an advanced degree.

But how can you know when you have enough nuggets and that they are suitably malleable? Almost immediately after asking yourself, Where do I begin? you should ask, Where do I stop? or How will I know when I'm finished? This can be a daunting question because humanity's quest for knowledge is boundless, the paths never end, and directions seem chaotic. Finding the answer is a process: you must clearly define your goal and objectives, plan your work carefully, study the literature, seek outside advice, and challenge yourself along the way, all the while learning the fundamentals of your discipline.

While the path leads to an advanced degree, it also extends beyond, toward intellectual maturity, along a route that climbs three levels of ascent: skills, technical knowledge, and scholarly thinking. Attaining skills and technical knowledge requires intensive effort, but the processes are fairly straight-forward—you study, you practice, and you learn. The third level, that of scholarly thinking and intellectual process, is more abstract and difficult. It requires not only that you be able to gather information, consider ideas, and make judgments, but also that you understand and control how you do these things—you have to think about *how* you think. Without this third step, you can do well, but you'll never reach intellectual maturity.

With determination, intelligence, and a constructive attitude, you eventu-ally will see vague beginnings progress to definite, valuable ends. But the progression is not spontaneous; it requires planning and a strong sense of purpose and process. In the end, your knowledge of process—of how to attack a problem and solve it—will be your main stock-in-trade. And it will serve you well for life.

First, You Need Advisors—and They Need Students
Selecting an Advisor

Your academic advisor, or major professor, should be your mentor and closest academic ally. He or she should be knowledgeable about your subject matter and enthusiastic about working with you. A good advisor will do all of the following and more:

- help you select graduate committee members
- chair your graduate committee
- advise you on courses to take
- inform you of departmental policies not covered in the university catalog
- help you select a research topic that has a high probability of success
- help you acquire funding and show you how to do it when you're on your own
- discuss expectations
- help you set deadlines and time tables

- meet frequently to evaluate and assist with your progress
- read what you write and critique your work
- challenge and guide you
- help prepare you for qualifying examinations and for your thesis defense
- perhaps lead a research group and help you and others in the group coordinate your efforts
- take you to professional meetings and introduce you around
- help prepare you to present your own work at professional meetings
- help you get your work published in a professional journal
- promote you in going on after graduation

It's a good idea to seek out and select your desired major professor before you apply for graduate school. Your choice of advisor probably will be more important than your choice of either a specific research area or even a university. Bear in mind that you must be able to trust and work with this person for several years. Choose thoughtfully.

Look for someone who will do all of the things listed above. (It's a big list—don't expect perfection.) Do not apply to work with someone simply because he or she has a big reputation or appears friendly and nice. Although reputation can be important, and friendship might develop, you first should look for a mentor: someone who will encourage and support your education and development, someone who will offer wise counsel, challenge you, and extend honest constructive criticism. Look for someone who will stick with you and your project. Beware of extremes: those who offer little beyond praise (especially as it pertains to their own work) can be as detrimental as those who criticize excessively.

Good mentors are proficient and respected in their fields, but more importantly they dedicate substantial time and effort to their students' development. The best mentors appreciate that their professional and social contributions subsist not so much in their own recognition as in the future contributions of their students. They want their students to be better than they have been, while concurrently they understand and accept that an individual student might not have that aspiration.

By now you might be asking, How do I locate potential advisors? Here are a few things you can do:

- Scan the recent literature (e.g., textbooks, scientific journals, economic reports) in your intended field. You'll find this in a university library and on the internet. (Caution: Don't rely on the internet entirely; go to a library, where you can actually touch journals and thumb through them.) Watch for current research topics and the names and affiliations of authors. At this time, don't worry about whether or not you can understand the articles—if you could, you wouldn't need an advanced degree in that field.

- Look beyond the university at which you earned (or will earn) your baccalaureate or first advanced degree. Check the catalogs and websites of several universities.

- Seek advice from faculty and other students at your present university. Although faculty must be careful not to overstep ethical bounds by publicly evaluating their professional associates, they at least can offer insights about who's doing what in the field, and they might tell you about people who are especially successful in securing research funding, attracting graduate students, and publishing papers. They will know prospective advisors at other campuses as well as at their own.

- Attend professional meetings. These might be local, regional, national, or international. Often, the details of time, place, program, and costs are given on the hosting organization's website. If you are unfamiliar with professional organizations, now is the time to learn.

- Do your homework, then visit a campus. If you hope to meet people, contact them first—don't simply drop in.

- Once you're on the campus, visit the labs that might interest you. Stop by the graduate student offices and the break room. Locate grad students who've been around a while—they'll tell you things no one else will. (But bear in mind that what they tell you is as much a reflection of themselves as of the professor.) While you're at it, stop by for a chat with the secretary or administrative assistant; just don't ask any direct questions about the faculty.

Once you've identified a few possible advisors, find out some things about them and their working group:

- What is her or his reputation, personally and as a researcher?

- How well does the professor support graduate students?

- How many graduate students does she or he carry?

- Are the students funded?

- What's the students' average time to graduation? How many have dropped out?

- How much are students expected to assist in teaching or work on projects other than their own?

- What's the professor's mentoring style?

- What's the atmosphere in the research group? Do students and post-doctoral fellows in the group work together and communicate well or do people stick to themselves?

- What are the professor's feelings about students beginning a doctoral program directly from a bachelor's degree program? (Earning a master's degree before a Ph.D. has some real advantages for the student.)

Don't forget, the process is not all about the advisor—it's about you, too. Ask yourself about yourself. What kind of person are you? Are you ready for the commitment? If you have a family, are they ready for the commitment? How

It's time for introspection: "What kind of person am I?"

much time are you willing to commit? Do you prefer guidance and exchanging ideas, or do you prefer to work alone? How well do you work with eccentric people? Are you eccentric? How much emotional support and pressure do you want? Are you adaptable? How thick is your skin? What would you bring to the atmosphere of the research group?

Inducing an Advisor to Select You

While you are looking for the ideal advisor, faculty are looking for the ideal student. Obviously, as in any successful relationship, your choice of advisor is not yours alone. You submit an application, which the faculty either accepts or rejects. Your advisor chooses you. So this is a good time for some introspection and self-evaluation. Examine your likes and dislikes, your fears and inspirations, your motivations and dissuasions, and most importantly your attitude. Examine as many aspects of your attitude as you can handle. Write them down (no one has to see what you write) and consider how they have shaped your past, influence your present, and presage your future. Think about changes you might want to make.

Faculty look first for a student who has a superior grade-point average, especially in the most recent two years of academic work. They scrutinize the courses taken, knowing the reputation for "easy A's" in some types of courses and the comparative rigor of others. They also check the student's scores on the Graduate Record Examination (GRE). They read letters of recommendation and generally know how to read between the lines when necessary. They value personal contacts and recommendations from people they trust. They pay attention to a prospective student's letter of intent and statement of objectives, all the while alert for clues to the student's dedication, outlook, and attitude.

The faculty who make the best mentors recognize that in education, process is more important than product. In other words, for a student's development, the findings of a project are secondary to the process by which the student reaches those findings. The ultimate success rests heavily on the student's motivation and attitude; faculty tend to weight these characteristics more heavily than precise grade-point averages and test scores.

Prepare yourself for interviews with prospective advisors. In addition to getting ready for the usual questions about your background, interests, goals, and the like, work on understanding your personality and proclivities a bit better. Of course you don't have to reveal your inner secrets, but you should recognize and think through some things about yourself *for* yourself. Here are a few questions to get you started:

- What are your distinguishing qualities, and would you like to enhance or alter any of them?

- How would you describe various aspects of your personality? For example, are you outgoing or shy, venturesome or guarded, playful or solemn, self-confident or hesitant; or are you somewhere between, depending on the time and circumstances?

- Are you a self-starter or do you prefer some direction?

- Can you tolerate uncertainty, or do you like to know where you are and where you are going? Do you need to feel secure in the value and validity of your ideas, or can you run on the faith that you will figure it out?

- What are some skills you'd like to enhance or develop?

- What are your values and principles, and what are they worth to you? If you ever are pressured to sell one—and you will be—how would you determine the price?

Selecting a Graduate Committee

Once you've been accepted into a program and have a major professor, you'll need to assemble a graduate committee consisting of your major professor (who heads the committee) and two or more faculty. The committee's purpose is twofold: (i) to educate, mentor, and guide you along the degree path and (ii) to administer certain requirements and uphold program standards. In performing these duties, committee members will provide expertise and counsel; review and evaluate your thesis; examine your knowledge, skills, and performance; ensure that program requirements are met; and, ultimately, decide whether or not your work is worthy of an advanced degree. (Note that some universities require each graduate student to have two committees, one for advice and support, and the other for evaluation.)

Here are a few tips on selecting your committee (including your main advisor if you don't already have one):

- Ask the department head or program director to recommend all faculty who have the expertise, interest, and time to work with you.

- Meet the faculty. Contact each potential committee member, and ask for a 15-minute interview. At the interview, begin with a few words on why you are there, then immediately state that you are *not* there to ask him or her to serve on your committee. Your purpose is to establish a mutual awareness. You will begin asking people to serve after you have met them all.

- Be prepared to talk about your background and interests and to provide a copy of your résumé if needed (see Chapter 6).

- Ask about his or her expectations of graduate students and concepts of thesis research.

- Pay attention to what is said and how it is said. For example, does the person speak about ideas and the work of former students or more about himself or herself? Are current and former students spoken of favorably or negatively, and is information about other people relevant and appropriate?

- Be ready to leave at the end of 15 minutes.

- If a potential advisor appears cool toward you or your project interests, consider another professor or alter your goals.

- Ask your major professor to suggest possible committee members.

- Select faculty who get along with each other. Although the student grapevine will tell you much about this, don't trust everything you hear. Ask faculty directly, but be tactful.

- If possible, take a class from each of the most promising faculty.

- Consider only faculty who have job security. Usually these are full-time, tenured or tenure-track faculty, although qualified adjunct and emeritus professors might serve as well. Avoid lecturers and other part-timers. Check with the department head or program director first.

- Be cautious about selecting faculty who appear eccentric or weird. Although they might be competent scientists and teachers, they likely will have difficulty interacting with other committee members as well as with you.

- In making the final selection, remember that committee members who are challenging and compassionate will be more helpful than those who are passive or self-important.

Getting the Most from Your Committee

First, remember that your advisor and committee members must play two roles: one as mentor to you, and the other as protector of academic and professional standards. Your role is to apply yourself diligently, accepting that graduate school is a full-time commitment—and then some. For the relationship to work successfully, you and they must establish an attitude of cooperation. Keep in mind that they are on your side; they are not adversaries. Strive to accept their critiques and criticisms openly. Do your best not to personalize their feedback, remembering that their intent is not to belittle you; rather it is to help you excel, and they'll do that in the best way they know how.

At the first committee meeting, you'll go through the regular talking points about your proposed program and research, your background, and goals. The focus will be on you, but at some point you should shift the focus to them. Ask about their vision of the graduate program, their expectations of graduate students, and their individual roles on your committee. Try to get a sense of how you all can work together toward your common goal. (And be sure that the goal is identified and agreed upon!) Ask whether or not you will be free to call upon them individually for research and other guidance. Some faculty will welcome open cooperation, while others might be sensitive about being perceived as interfering with or taking over responsibilities of the major professor. Even if you think you've settled these matters individually before the meeting, go over them again as a group.

Don't hesitate to use your committee at other times. Schedule regular meetings with your advisor, perhaps once per week. Of course you won't need to meet every week, but both of you should be thinking about it. Communicate with each committee member at least once per academic term, and meet with your full committee once per year. Remind them about your project and your goals, inform them of your progress, and seek their insights. Take careful notes each time you meet, whether as a group or individually. Also, think about what you'd like to accomplish at the meetings, and prepare notes to take with you.

Finally, remember that graduate committee members almost never qualify for sainthood. Like the rest of us, they are afflicted with the hang-ups of humanity. One or another of them might be a sinner today and a savior tomorrow (or vice versa). In short, you should expect and demand the best of them, but not perfection.

Your Responsibilities to Your Advisor, Your Committee, and Your Group

Although you are the student, part of your job is to help the faculty do their jobs well. While you expect something from them, they should be able to expect something from you. A teacher cannot teach and a mentor cannot advise and encourage without assistance from the student. Most faculty probably would agree that the *three most important attributes of a good student are a positive attitude, motivation to learn and work, and willingness to communicate.* Even genius intelligence cannot make up for deficiencies in these three areas.

As a graduate student, you no longer are in a position of merely receiving an education. You are in a transition phase in which you are expected to give as well as receive. You must contribute intellectually by sharing your insights, knowledge, and intuitions with the faculty and your fellow students. They will ask your opinion, and you will be called on to critique their work—yes, even that of the faculty. In this regard, you will have become a junior colleague. Part of your job is to offer observations and recommendations that are thoughtful, forthright, respectful, and honest.

In meeting these responsibilities, as well as those to yourself, you must keep up to date in your chosen field—often even more up to date than the faculty, including your major professor. Although any good professor regularly will attend professional conferences and read current professional papers, his or her other duties (including keeping graduate students funded) leave little time to thoroughly scour the literature. That's your job. To accomplish this you must read, read, read, and then keep your group informed by sharing your latest findings. (See "Reading and Evaluating Research Reports" and "Preparing a Literature Review?" in Chapter 4 for hints on reading and maintaining a file.)

Your duties will extend beyond purely academic matters. If you work in a laboratory or are otherwise part of a research group, you must assume some responsibility for the proper functioning and success of the group. In addition to communicating and maintaining your best possible attitude, you will need to take on part of the laboratory management. You should willingly assist laboratory technicians, post-doctoral fellows, and other students, whether their experience and knowledge are greater or less than yours. (Normally, a doctoral student assumes greater responsibility than does a master's student.) An important reason for taking responsibility now is to prepare you for future leadership.

Here are a few more pointers, including some additional responsibilities:

- Take initiative, accept jobs, and do the chores—sometimes even if they don't pertain directly to your work. When something breaks, fix it; when the lab gets messy, clean it; when the printer ink runs low, replace it. You get the idea.

- When a new student joins the group, help her or him out. Take a walk to the library, student center, gym, and bookstore. Try to make the newcomer's first experience as positive as possible.

- Learn when and how to delegate. That's easy to say, but how can one develop the skills? Begin by observing someone who is good at it. Note how this person interacts. You'll find that a good delegator knows his or her relationship to others, understands what needs to be done, gives clear instructions, and follows through.

- If something is troubling you, tell your advisor and resolve the problem directly. Don't simply complain to other students.

- If a problem seems insurmountable, or a committee member seems unapproachable, tactfully seek advice from another member or the university's counseling center. Don't brood about it and continue complaining to other students.

- Accept that graduate school is more than academics and research. It's also about learning to work together and be productive in a demanding environment, whether the group members are fully compatible or not.

- Attend professional conferences with your major professor and committee members. Join and participate in local and national professional societies.

- Write out your academic and professional goals (see "Academic and Professional Goals and Objectives," later in this chapter). Attach these to your résumé, and present to each committee member.

- Write your research thoughts and goals as they evolve. Periodically arrange them into a brief proposal (including title), and present it to each committee member, stressing that these thoughts-in-progress are subject to change. This preliminary design can evolve into your research plan (see Chapter 3). Be sure to date each version.

- Don't circulate anything of substance to your committee members without first consulting your advisor.

- Plan on some long hours and hard work—50 to 60 hours per week (in addition to classes) is common in the sciences.

- Learn about the budget and operate within it.

Next, You Need a Research Topic and a Strategy

The process of selecting a research topic varies among graduate programs and even among students who study under professors in the same program. While some students might be assigned a major professor and an ongoing project at the time of their acceptance, others are left to find their own path. Still others fall between the two extremes: they might be assigned an advisor who has expertise in a particular discipline, but then they are expected to determine their own focus within an overall research mission. Students who have the greatest difficulty seem to be those who set their minds on a specific topic, then search for a mentor to fit it. In any case, master's-degree students normally are given greater guidance than are doctoral students.

No matter what your situation, read through the following sections.

Criteria for Identifying a Suitable Topic

Topics for study are ubiquitous and limitless. The challenge is finding one that is relevant, researchable, and right for you. These three Rs sound like they could head a criteria list, which they do (even though they're not necessarily presented in a priority sequence):

- The topic should be relevant. To be relevant, a graduate research project must be important to someone beyond yourself. Some part of a society should be able to benefit from your efforts, although the timing and sometimes the exact nature of the benefit might be left unspecified (as in basic research). To establish importance, you must read the literature and interview knowledgeable people.

- The topic must be researchable. A researchable topic is one for which you can set realistic goals and envision a successful end. The project can be achieved within the limits of your capabilities, resources, and time. Your intellectual capabilities are determined by your ability to perceive and comprehend, combined with your determination; they are not set by your knowledge and skills at the time you begin. To start, you need background knowledge in the field of your topic; developing advanced understanding is a main purpose of pursuing the degree.

- Look to your graduate committee for help in tapping resources. Every research project needs funding and logistical support, regardless of the sources.

- Two years beyond the bachelor's degree is reasonable for completion of a master's degree, and three to four more years for a doctorate. More time might be needed if you change fields from one degree to the next.

- Again, look to the scientific literature for help in determining whether or not a topic is achievable.

- It should be right for you. Does the project seem interesting? Is it attractive? Does it have any aesthetic appeal? Does it fit with your professional goals? Obviously, it should, but don't let an initial lukewarm interest deter you from considering a potential project. Usually, a student's interest grows with involvement, as a feeling of ownership develops in the project.

- In addition to feeling right for you, your project must be acceptable to those who will guide you, fund you, and judge your academic performance. These individuals or groups include your graduate committee, the university's research administrator, the funding agency, and perhaps even a client or employer if they provide support. If your committee members are not enthusiastic, find another project, another committee, or even another program.

- The problem to be investigated should be unsolved. Again, you must read the pertinent literature and interview knowledgeable people.

- Finally, remember that your number-one purpose should be to complete your degree requirements. Your thesis work is not an end in itself, rather it is a means to an end that is defined by your long-term personal and professional goals.

Recognizing and Selecting a Researchable Goal or Problem

Selecting a suitable problem or goal is fundamental to all worthwhile research. It requires that you be curious and observant and that you cultivate your capacities for passion and compassion. Pragmatically, you instigate your curiosity, compassion, and observational skills to identify your clients and recognize their needs. Research "clients" are not named individuals; rather, they are any group that might benefit from or support your work (e.g., resource managers, investment bankers, cancer patients, potato growers, government agencies, other researchers, benefactors).

You can find research needs almost anywhere, including textbooks, scientific journals, government reports, newspapers, broadcasts, popular magazines, and internet websites, as well as through personal communications. One problem you will find is that the needs rarely will be presented as a researchable question. Defining the question is your job. Here are a few pointers to get you started:

- Visit with knowledgeable scientists and practitioners.

- Amass background material, and look for gaps in knowledge and inconsistencies in viewpoints. Be aware that some sources, including textbooks, typically cover over gaps and discrepancies to make a flowing, coherent story. Exercise caution with internet sites—many are not trustworthy.

  (Note: If you are a master's-degree student, the full job of seeking out and learning the literature will require a good six months; doctoral students should plan on a year. Right now, you need to do only enough to get started.)

- Look for analogies to related problems.

- Conduct local field work to collect first-hand information.

- Find out where the money is; then go after it, bearing in mind that you might have to adapt yourself to grant requirements. Consult with your advisor and committee members, and perhaps with the folks at your university's grants development office. (Don't consult the grants office without your advisor's approval.)

- Think in terms of topics and overall problems at first, then brainstorm to identify subproblems and critical questions relating to the overall problem.

- Focus on problems that are solvable and goals that are achievable. Note that this advice does not mean that scientists should shy away from risk; no progress would be made if they did. Nor does it mean that you should not take intellectual risks during the course of your work; you should. Rather it means that in the beginning you should set yourself up for success—no prizes are given for failure to solve a problem.

- Maintain a research notebook and write your thoughts and progress in it.

- When you think you have found the perfect problem, poke a few holes in it before someone else does that for you.

- To be worthy of thesis research, a project must be more than a simple exercise in gathering data; it must be more than a simple comparison; it must require more than a simple yes or no answer; and it must go beyond a simple correlation between variables. It must have an analytical component (See "A Definition of Thesis," presented in Chapter 2.).

Finally, in searching out your topic and problem or goal, consider the criteria presented in this and the previous section. At the same time recognize that science should not be limited to industrial thought. It should not devalue intuition and the benefits derived from "right-brain" inspirations. So heed your intuition and follow your inspiration—at times these can reveal more than cold, logical analysis. Still, you will need logical analysis to challenge and harness intuition and to take you beyond its capabilities.

Setting Working Goals and Leaving Tracks

Once you get going, you will need to set working and accomplishment goals designed to help you make day-to-day progress. Without these goals, the chore of planning, doing, and writing a quality graduate thesis will be overwhelming. Separate your thesis work into manageable parts, and then divide those parts into smaller parts. The various sections of the research plan outline presented in Chapter 3 of this manual are good first breaks.

In defining smaller, more specific goals and objectives, set yourself up for success and frequent rewards. Set an objective for each working session, with each one having a clear beginning and definite end. Begin with what you can do. Here are some examples:

- Learn to use the library. (Do not assume that the internet has rendered libraries and librarians obsolete!) Become familiar with books in your field, locate current periodicals, and browse through a few scientific journals. Also find out about scientific abstracts, maneuver through government documents, and introduce yourself to the librarians.

- Read and make notes on one research article.

- Write 400 words of literature review.

- Set up the thesis title page, table of contents, and list of figures.

- Format three tables and one figure.

- Plan the time table to include in your research plan.

- Statistically analyze one data set.

- Outline major points for the discussion section.

At the end of each session, write notes that clearly show where you left off and instructing you where to begin in your next session. Outline the jobs you plan to accomplish next time. Give yourself lots of directions. Bear in mind that each time you complete one of these objectives, you have moved a step closer to completion. Smile and feel good about it.

Some Personal and Practical Points

On Being a Graduate Student

A person's progression from elementary school through high school and college is a series of successive stages in the individual's formal education. Normally, the K–12 years are a continuous sequence of increasing experience, knowledge, and personal responsibility, with the length of steps increasing only moderately from one year to the next. Then, from high school to college, the step is longer than any taken before, and, if one pursues a research-oriented advanced degree, the leap from undergraduate to graduate student can be even greater.

Your transition to graduate status is more a leap than a step for at least three reasons. First, your depth of learning and the expectations placed on you intensify substantially. You are expected to have superior motivation and study habits as well as adequate fundamental knowledge to undertake advanced studies. Your scholarship requirement, as reflected by a minimum grade point average, is higher as a graduate student than as an undergraduate. Also, you are expected to have advanced maturity and an attitude that displays openness, eagerness, and dedication, all of which are needed to successfully assume your greater responsibilities.

Second, when you transform into a graduate student, your relationship with the faculty changes. Your teachers no longer are superior beings who simply direct and explain. Rather they are mentors who suggest and guide—and in some measure learn with you. Under these conditions, your approach to scholarship and your mechanisms for learning must adjust accordingly. You must become more self-sufficient, self-regulating, and self-motivating. You must become more independently discerning and challenging of yourself as well as of others.

Third, and perhaps most significant, graduate school is far less structured than anything you've probably been used to. Rather than other people telling you what to do, you have to figure out much of it yourself—then do it with commitment. You are expected to create your own structure, especially at the doctoral level. Of course your advisor should lend support, but the bulk of the responsibility is yours. You must shape your program, locate and read scientific papers, design and carry out your research, and write your thesis—all without anyone telling you to do these things. Fortunately, you'll be among a group of self-motivated, dedicated people who love to work and who mostly are eager to help each other.

While you are gaining knowledge and discipline, you also will have opportunities to learn new skills. Don't shy away from opportunities. Whereas some of us prefer to amass facts and contemplate ideas, others would rather develop proficiencies and invent techniques. Although these tendencies reflect our temperaments and backgrounds, they also can inhibit our development. So be wary of letting your predilections control your education. If you enjoy reading and thinking, make time to learn and practice skills (e.g., laboratory methods, analytical techniques, computer programs). If you tend toward technology, study the theory behind it. In short, broaden out, take a few risks, and march bravely toward new horizons—this probably is the best opportunity you ever will have.

Seek new horizons and march bravely toward them.

Communicating

Not many things, if any, in graduate school will cause you more trouble than a deficiency of open, honest, frequent, timely, and helpful communication with key people. A simple, albeit sometimes difficult, solution is at hand: communicate frequently, freely, and honestly with your major advisor, committee members, other faculty, and other students. Discuss your work and express your feelings about it. Although at first these conversations will point you in different and sometimes confusing directions, ultimately they will help you set your course, find your way, and strengthen your resolve.

Speak about your work with laypersons and professionals from other fields as well. Interested family members and personal friends can be good sounding boards. Be aware that you are on your way to becoming an expert in your subject; uninformed people will look to you for information, advice, and answers. Don't take the attitude that they wouldn't understand it; think of ways you can explain your work to them. Be aware also that often they won't know the proper question to ask. Part of your job will be to help them express it.

When talking about your work, consider the image you want to present. In large measure, your image will determine how your message is received by others. This point is especially important in front of an audience, which sooner or later is where you likely will find yourself. The most effective speakers project honesty and authority without arrogance. Is this the way you present yourself? Is your scientist's image different from your personal image? Should it be? Use of language is important. Practice using clear scientific terms and explanations; avoid jargon.

Frequent discussions with committee members and others can help you develop authority in the subject matter, but for your authority to be comprehensive and honest, you also must exercise the other aspects of communication: reading and writing. Read textbooks for overviews, fundamentals, and contexts. Read scientific journals for details, methods, and latest developments. Make notes on all the pertinent articles you read. Assemble these notes and write a comprehensive literature review—this can be one of the most fruitful exercises you undertake (see Chapter 4 in this manual). Remember that true communication, which fosters clarification and understanding, is active, not passive. Reading without writing is a passive endeavor—and so are talking without listening and listening without talking.

Academic and Professional Goals and Objectives

Where do I want to go? How do I want to get there?
What do I want to do? How do I want to do it?

To be successful, a graduate student must face these questions directly, but sometimes it's easier, and immediately more fruitful, to first express where you *don't* want to go and what you *don't* want to do. In either case, think about these questions carefully, pay attention to your inner feelings and intuition, and allow your decisions to be flexible but not erratic.

Your academic goals define what you want to achieve educationally. They might define the culmination of your formal education, or they might extend beyond. For many people, academic goals direct them *toward* a career; their professional goals direct them *through* that career.

Here are a few benefits of writing academic and professional goals and objectives:

- clarifying your thoughts
- helping you to select a thesis project
- helping your advisors and associates to help you
- strengthening your résumé and job applications

So what's the difference between a goal and an objective? Although the answer varies somewhat among authors, this manual makes the following distinction:

> **Goal:** a long-term end, an ultimate purpose, or an intended outcome toward which we direct our efforts. Goals help to guide our actions, even if they suggest a somewhat general, remote, or idealistic purpose.

> **Objective:** a specific subset of a goal, usually shorter term, more concise and concrete, and more immediately attainable, with no obviously insurmountable barriers.

You can have goals without objectives and objectives without goals, but you won't make much progress, except by muddling through.

Time Management

Although succeeding in a rigorous graduate program can return lifelong rewards, the route to success can be as tiring as it is daunting. Graduate students who stay with the program quickly learn that time management is critical, not only for their success, but also for their well being and happiness along the way. As a graduate student, you will need to balance time for a great variety of demands, which will include most, if not all, of the following: course studies, seminars, grant-proposal writing, laboratory and field investigations, data interpretation, library study, report writing, teaching, and, oh yes, home life. Sleeping and eating along with a little exercise and recreation need to be squeezed in as well.

The better you are at scheduling time, setting priorities, and postponing procrastination, the easier your life will be. You can find numerous self-help books on the subject, and perhaps they have some good suggestions. Here are a few more:

- Try to emulate behavioral patterns and techniques of people who always seem to be on top of things.

- Set small (i.e., achievable) daily goals.

- Get things ready the night before.

- Schedule the most creatively demanding work for the morning (but not for mornings when you've stayed up all night).

- Start assignments on the day they are assigned.

- Work on assignments and projects a little at a time, rather than in one long sitting.

- Keep a calendar of your activities and deadlines.

- Maintain an academic notebook and a personal diary.

- Write as you go along.

- Make notes as you read scientific papers.

- Avoid scheduling course work, research, and assistant teaching in the same academic term.

- Back up your computer files daily, or you'll spend considerable time redoing work.

Money

Graduate students are always broke—not really, but that's the way it can feel sometimes. As a graduate student, you probably will need to budget your money carefully, but first you'll need an income. Any university with a good graduate program will offer a variety of financial aid opportunities, ranging from student loans to free-ride fellowships with a stipend. In other words, you can get paid to go to school. You should inquire about funding possibilities before you apply to a program. Usually, these are posted on the university's website and in broader internet databases. When you arrive on campus, visit the financial aid office.

Financial support normally is offered in several forms, including assistantships, fellowships, grants, and scholarships. Here's an overview of each.

Assistantships normally are offered in two forms, teaching and research. With either form, the student receives payment with fringe benefits (e.g., health insurance) and usually a partial or full waiver of tuition and fees. The student performs services while continuing normal studies and maintaining acceptable academic standing. The services, which benefit the student as well as the university, are learn-by-working activities in the student's area of interest. A student might be a teaching assistant one academic term and a research assistant another. Assistantships normally run year-round, including summer.

Students who are provided a teaching assistantship (TA) teach part of a course under the supervision of the faculty member who has primary responsibility for the course. Usually, the professor delivers the lectures and the TA instructs laboratory or discussion sessions. The TA also gets the pleasure of grading exams and papers. The job commonly is based on a 15–20-hour work week, but hours can vary.

Students who are awarded a research assistantship (RA) conduct research that most often is their thesis work, although additional part-time duties might be assigned. Working arrangements, including the expected working hours for an RA, can vary greatly.

Fellowships, which are less common than assistantships, are outright financial grants to exceptional students. They normally provide a nine-month or full-year stipend, benefits, tuition and fee waivers, and a book allowance. Although the student has fewer direct responsibilities and more working freedom than with an assistantship, the student still is expected to perform a service by conducting research and publishing. The student must maintain high scholarship and is expected to bring honor to the university. Fellowships are offered by outside funding organizations (e.g., foundations, governments, professional societies, and private endowments) as well as by universities.

Grants usually are more directed and goal-oriented than fellowships. They are awarded by governments, foundations, and private companies for specific purposes. In addition to funding for students and faculty, grant budgets include all other costs of conducting research. They normally require progress reports and sometimes are renegotiated periodically.

Scholarships mostly are one-time awards that vary from hundreds to thousands of dollars. They are endowed by numerous types of organizations and even by private individuals. Although students must meet specified criteria to be eligible, they usually are free to spend the money as they see fit.

In nearly all cases of financial aid, you won't get it if you don't ask for it. Don't be shy—apply! You will have considerable competition for some opportunities but surprisingly little for others. Generally, the higher your grade-point average, the better your chances of being funded. But don't let a less than perfect grade point discourage you from trying. You rarely will know where you stand against your competition or how the awards committee views the various applications. So go for it!

Finally on the subject of money, here are a few additional random thoughts you might find helpful:

- Set yourself a budget and live within it. You'll have enough intellectual stress as a graduate student; don't pile financial stress on top of it.

- Beware of building up debt, especially that incurred by using credit cards. If you can't pay off the credit card each month, don't use it.

- Consider your idea of a luxury, then reconsider it. Luxuries are relative to our life-styles and attitudes. Many luxuries that can be important and satisfying in graduate school cost very little.

- Check on the tax requirements related to the various forms of financial aid. Some are tax free, others are not; and the tax laws change frequently.

- Even after you're settled in and have acquired funding, continue to check with your financial aid office periodically for possible new opportunities. Also check online periodically by typing "graduate scholarships" and related terms into your favorite internet search engine.

- Once you've been awarded funds, don't spend wildly, thinking that this income will last for as long as you're in school. It might not because many awards, including assistantships, are supported by "soft" money, that is, money from ephemeral, albeit renewable, sources. These awards depend on availability of funds.

Ups and Downs

Any reputable graduate program will be challenging, stressful, and rewarding, and as a serious student, you probably will experience swings in your mental state and productivity. With your attention focused intently on your thesis, your life sometimes will be marked by antithesis. One month you might sparkle with enthusiasm, motivation, cheerfulness, and self-confidence, and the next you might stultify with disinterest, frustration, anger, and depression. These swings can range from slight to severe, and mostly they are common and natural—to be expected. Nearly everyone goes through them.

Although sporadic surges and ebbs that punctuate your progress and disrupt your psychic equilibrium can be unsettling and even unnerving, they nonetheless are necessary for intellectual development. Without them your work would become inanimate and uninspired, even dull. You can't develop intellectually without them.

While riding waves of emotion and spirit, you probably will find that your inspiration and productivity will come in fits and starts. You will have up times and down times. Try to remember that even when you are in a low, the wave moves forward. Some part of your mind will be cogitating—preparing to ascend the next swell—and sooner or later you will move forward.

Knowing that you should expect times of intellectual and emotional taxation can help you endure them and come through stronger. Although these periods can feel repressive, they do not have to be immobilizing. These can be good times to monitor your feelings—to look into yourself, preferably in ways that an honest, objective, and sympathetic outsider might see your inner being. With effort and practice, you can learn to cultivate and fortify your emotions, and eventually to manage the ways in which they drive your motivation, inspiration, and creativity. (See Averill and Nunley, 1992 for some helpful insights.)

Although it's an important part of the graduate-student experience, learning to expand, manage, and utilize emotions is a lifelong process. If you're a graduate student headed into a slump, you need more immediate solutions. Here are a few suggestions for survival:

- Prepare for the slumps; they can appear without warning. For this, you need a clear view of where you have been, where you are, and where you hope to go. Be sure that you know and understand the fundamentals underlying your discipline and your research. Prepare a comprehensive written research plan that includes a thorough review of appropriate scientific literature (see Chapters 3 and 4). Record details of your work as you proceed (see Chapter 5).

- Prepare for the slumps also by recognizing and accepting that despite the need to have a clear view, your vision frequently will be muddied. Graduate school is fraught with uncertainties, and times will come when these will be upsetting. From time to time, you will doubt your work, your ideas, and even your personal goals and abilities. That's normal—it goes with the territory. After all, you can't expect to investigate the unknown, including that within yourself, without encountering ambiguity, doubt, and indecision.

- Sometimes that uncertainty will seem to creep into your soul as well as into your research questions. When it does, guard against grasping too quickly at a solution that gives speedy relief at the price of longer-term misery or embarrassment. Learning to tolerate and cope with uncertainty can be difficult, but it also can be one of the greatest rewards of graduate school. Ultimately you will develop greater faith in yourself and your abilities. You will learn to operate on a higher plane, and you will come out stronger on the other side.

The way it sometimes feels.

The way it is.

The other side.

Still water can be relaxing, but it won't take you anywhere.

- Soften the impact by hardening your body. Exercise regularly to the best of your physical ability. A light workout for 15 minutes a day is time well spent. Exercise strenuously three times per week. This is one of the best ways to alleviate anxiety. (If you feel really down or are facing a big exam, avoid competitive exercising such as tennis. Missing a shot will add to your frustration. Stick to calisthenics, running, swimming, and the like.)

- Lay off caffeine, alcohol, and nicotine, especially when you have a deadline approaching. Although these and other drugs might stimulate poetic and fictional creations, they can stifle scientific creativity and productivity.

- Keep your goals and perspective realistic. Having reasonably achievable goals can help you avoid a fear of failure (or for some of us, a fear of success). Remember: "Some people suffer from an exaggerated fear of failure because they have an exaggerated view of success" (Ladd, 1987, p. 41).

- Compartmentalize the jobs you need to accomplish and work on those that require the least thought. For example, you might set up your thesis title page and references section, repair field equipment, or wash glassware. You can make progress even when you feel as though you're not.

- Work on compartmentalizing your thoughts and set a time limit for entertaining the discouraging ones (15 minutes per session is reasonable).

- Don't try to suppress your thoughts, moods, and emotions. Instead face them, examine them, and write them in a diary. Be honest in your writing—you don't have to show it to anyone.

- When you anticipate an impending problem, try to visualize yourself handling it successfully—before it arises. Don't tie yourself in knots trying to second-guess or outmaneuver another person, but do think about how you should respond in challenging situations. Avoid win-or-lose confrontations; in the long run, these usually are counterproductive.

- Avoid feelings of isolation. Maintain contacts with friends, family, your committee members. Don't bother trying to cover up your frustrations when interacting with others, but don't belabor them either. (It's OK to wallow in your misery for short periods. Set a time limit and tell people close to you what you are doing.)

- Schedule fun times and social engagements.

- Read a good novel (or a bad one if you want). And while you're engaged in diversion reading, peruse some haiku—especially that written by children. Try writing some yourself. And check out Dr. Seuss's *Oh, the Places You'll Go.*

- Visit your campus counseling center, and if the first counselor you meet is not right for you, ask for another.

- And sometimes you just have to abandon whatever you are doing. Take a break and come back later. Don't burn bridges, and be sure to inform your committee members and others close to you what's going on. Then go.

Stress-Strain Relationships

If everyone—you and each of your committee members—is well adjusted, perfectly honest, altruistic, understanding, and reasonable all of the time, all should be fine. If not, you can expect occasional tension or perhaps even conflict. This goes with life.

Being a graduate student can be stressful, as can being an advisor to graduate students. Stress, according to physics books, is a directed pressure. That means it's an imposing force that comes from a usually identifiable source. Stress causes strain, which means that things become deformed; they get pushed out of shape. The greater the stress, the greater the strain.

A strained relationship can develop when you and your advisor are under pressure. These times call for extra tolerance, understanding, and communication.

Physically, stress can pass across a contact from one object to another. Meta-physically, stress can pass between you and your advisor, causing a strained relationship. The complexities of stress and strain in human relations are great and best left to psychology. Nonetheless a fundamental coping strategy can be outlined.

The strategy for handling stress of this type can be summarized in five choices: (i) suffer it, (ii) prevent it or deflect it, (iii) block it or decrease it, (iv) strengthen your resistance, and (v) remove yourself from its path. You'll have to figure out which mechanism or combination suits your situation and circumstances. Here's a bit more explanation:

Suffer It

You should be able to endure misunderstandings and minor disagreements. If kept in perspective and addressed promptly and forthrightly, these will be inconsequential. But you should not allow yourself to suffer under continued irritation. That breeds frustration and contention. The amount you choose to suffer is up to you. Perhaps you know your tolerance level, but perhaps you don't. Work at monitoring your feelings and your responses to comments and events that appear significant. Record them in your diary and watch for patterns that might suggest further direction.

Prevent It or Decrease It

Try to head off or resolve misunderstandings and disagreements before they become stressful. This is partly your advisor's responsibility, but it's also yours. If you detect a potential problem, bring it to the fore. No one can discern your feelings and concerns telepathically. You must do your part to be as open as you can and not allow minor difficulties to breed a dysfunctional relationship.

Strive to recognize interpersonal differences and seek an agreement for tolerating them. This approach can be difficult because it requires mutual honesty, respect, and trust. Adjustments must be made at the source, which could be your advisor, you, or both of you. When successful, this approach is among the most rewarding and long-lasting.

Take some time to carefully examine your own actions and how they might affect other people. Sometimes a person who is the recipient of undue stress also can be the cause. If things that we do, or should do but avoid, initiates stress in another person, that stress can be reflected back to us.

Don't look for problems without due cause, and don't try to capitalize on differences or disagreements.

Deflect It or Dissipate It

This approach requires a shield or a buffer—usually another person who can redirect or help you absorb or disperse the stress. If your advisor has proven to be unreceptive, you should seek help from an advisor in your departmental graduate studies office or campus counseling center. If the situation remains unresolved, you can contact the faculty member's supervisor, who usually is the department head. Try always to resolve the problem first personally, then at the lowest possible administrative level. Before expressing your concerns in any office, ask about confidentiality policies. Then be as tactful and fair as you can.

Undoubtedly, you will talk with students and others you trust—but watch yourself, your purpose should not be simply to gripe or bash your advisor. You should seek insights that can help you understand and deal with the problem.

Strengthen Your Resistance

Difficult though it might be at times, do your best not to take adverse comments personally. These usually are addressed more at an action or situation than at a person. Unfortunately, they're not always properly expressed that way, and hurt feelings are a normal reaction. Still, being aware can help.

Also be aware that your hurt feelings might indicate an overly sensitive reaction and that you are taking personally something that has no personal basis. If this is the case, you might be trying to cope with a stress that is more internally than externally imposed. You'll have to look into yourself to deal with this one. But don't beat yourself up—it's a common response among thoughtful people.

Remove Yourself from Its Path

This can be the most drastic of the choices, because it could mean changing advisors or quitting the program. Usually your best course is to exhaust other potential solutions first.

Here are a few additional points and suggestions:

- If seemingly inconsequential negative events recur even after resolution attempts, they could be a sign of a different, perhaps larger problem beneath the surface. You'll have to monitor the situation, and if possible try to discuss it with your advisor or a counselor.

- Cut your advisor some slack. Remember that having a doctoral degree means that the person has been highly educated and trained in a particular field. It doesn't mean that she or he has attained infinite wisdom or the ability to work effectively with others. Nor does attainment of the degree liberate one from health problems, family squabbles, and other anxieties and demands of everyday life. Your advisor could be trying as hard as you are to cope.

- Always be courteous and respectful, bearing in mind that communication is more than words—it's also how the words are delivered and received.

- Read the relevant thought tools (i.e., dualities) for problem solving in Chapter 2.

Finally, consider this: within the next few years, you could be in an advising role. What will you have wanted to learn from your experience as a student? How would you prefer your students (or employees) help you do a better job?

Advice on the Mundane Side

- Read the catalog requirements and know the proper name of your intended degree before requesting initial interviews with potential committee members.

- Many programs provide a graduate student's handbook that will tell you about selecting a committee, taking courses, getting a parking permit, using the health center, and a whole lot more mundane, but useful, information. It's normally on the internet. Check it out.

- Learn your computer programs, most importantly word processing, database, statistics, and visual presentation.

- Use the library; don't rely on the internet for everything. Use the library and the librarian. Librarians are helpful people.

- Attend other students' oral examinations and theses defenses, and be sure to invite other students to attend yours (check with your advisor first).

- Join and actively participate in a prominent professional society in your field. These commonly are at various levels, ranging from local to international. You will meet stimulating people and advance your career while supporting your profession. And believe it or not, established professionals will want to meet you.

- Accept that the process of completing a thesis will require more time and harder work than you think. Unless you're in a weak program, it always does.

Major Hurdles: Preliminary Exams and Final Defense
Preliminary or Qualifying Examinations

What Are They?

A few master's degree programs and all reputable Ph.D. programs require students to pass a series of written and oral examinations early in their studies. For master's students, these exams might be administered at the end of the first academic term, and for doctoral students, the exams normally take place within the first two years of study, and in some programs before the student is allowed to begin research.

These exams ascertain competency and identify weaknesses in the student's preparation for the advanced degree. They test basic knowledge and analytical abilities, as well as perhaps technical skills in the discipline. Normally, master's degree students must prove excellence at the highest undergraduate level, while doctoral students must demonstrate ample expertise to teach senior-level undergraduate courses. Compared with master's students, doctoral students must show a more developed aptitude for critical and abstract thinking, as well as greater ability to work independently.

The results of the examinations guide faculty in judging whether or not a student is sufficiently dedicated and prepared to remain in the program. If the results are positive, the student will be advanced to candidacy, a status necessary to further pursue the degree. If results reveal deficiencies, the student is either given direction for improvement and reexamination or released from the program.

What to Expect

Examination formats and performance standards vary greatly among universities and somewhat even among programs within a university. The structure might be rigid or loose; subject areas might be designated or open-ended, and the time allowed can extend from a few hours for master's exams to two weeks for Ph.D. exams. Questions, which can be closed-book or open-book, might be segregated into sections, with passing of the first section required before the student is permitted to take subsequent sections.

Questions are submitted from a faculty committee, which normally is broader than, or sometimes different from, the dissertation advising committee. In some programs, the entire departmental faculty and a member or two from a related department will participate.

In addition to directly testing technical knowledge, many doctoral examinations require students to prepare a research proposal, which might or might not relate to their discipline. The purpose is to assess the student's potential for analyzing problems and proposing solutions as well as to evaluate his or her immediate knowledge of scientific thinking and process. An added value to the student is that he or she gains experience in meeting a major career challenge. At most research universities, the student may prepare by taking a course in writing research proposals and securing grant money. (Chapter 3 in this manual is devoted to writing research plans.)

To learn more about the structure and requirements of examinations in your program, consult your major professor as well as your university's catalog and webpage.

How to Prepare

First, consult your major professor and other committee members, then talk with other students who have been through the process. Probably, a file of past exam questions exists somewhere; perhaps it's even posted on the internet. Although your questions likely will be different, old questions can give you an idea of what to expect.

Most importantly, be sure you know the fundamentals of your discipline. Read a comprehensive introductory text and your old course notes, whether they are from the same university or a different one. If you haven't assisted in teaching the main introductory course, now's the time to do it—teaching is one of the best ways of ensuring that you understand a subject.

Be familiar with major journals in your field and read as many papers as you can. Pay particular attention to classic articles, recent advancements, and papers published by faculty who might serve on your testing committee. Make notes on what you read.

Possible Outcomes and Consequences

Performance standards, evaluation criteria, and possible outcomes vary—check with your major professor. Usually, qualifying exams are graded in a pass-fail system, with provisions for intermediate rankings (e.g., pass, conditional or provisional pass, fail). A passing grade allows the student to be advanced to candidacy and continue in the program. A conditional pass usually means that the student must complete additional preparation and retake the examination or some part of it. A failing performance usually results in the student being dismissed from the program.

In some cases, a student holding a bachelor's degree might have been admitted directly into a Ph.D. program without first earning a master's degree. A student in this status who performs below acceptable doctoral examination standards might be allowed to continue toward the master's degree but not toward the Ph.D.

The Final Defense

Typically, the concluding requirement for a research-oriented advanced degree would be for you to successfully defend your thesis or dissertation. The examination has three purposes: (i) to test your knowledge and thinking skills, (ii) to give you a chance to show off your work, and (iii) to help you prepare for future challenges. The defense commonly consists of an open, one-hour presentation to faculty and students followed by a closed, one- or two-hour oral examination before the graduate committee. The results of this defense determine whether or not you have fulfilled all requirements for the degree.

Although examinations normally emphasize research work and closely related topics, their formats and performance requirements can differ among programs. Your major professor should clarify what to expect several months in advance.

Here are a few pointers that can help you *prepare* for your oral defense:

- Know the fundamentals of your discipline as well as the underlying principles and technical aspects of your research work.

- Have an up-to-date grasp of the literature in your field (see the section in Chapter 4 "Preparing a Literature Review").

- Know the immediate utility and broader potential applications of your research.

- Be prepared to explain why you chose to proceed one way or another during the course of your work.

- Recognize the limitations and shortcomings of your research and consider what you might have done differently. Be ready to suggest possible future work for yourself or other students.

- Expect questions that go beyond the scope of your work—questions that test your ability to "think on your feet." A firm understanding of the fundaments and process covered in Chapters 2 and 3 will help here.

- Know why you chose to pursue an advanced degree and what you hope to do with it.

Keep these points in mind *during* your oral defense:

- Remain as upbeat and positive as you can. Projecting enthusiasm and pride in your work will help put you in control of the situation.

- Do your best not to internalize or take personally any question or criticism from any committee member. Bear in mind that faculty sometimes will try to provoke discussion and prepare you for future confrontations by playing the devil's advocate.

- Make notes as questions are asked. This will help stimulate and direct your thinking, as well as ensure that you answer the question completely as asked.

- If you don't understand the wording or purpose of a question, say so.

- If you don't know the answer to a question, say so. Admit that you don't know, but then continue with how you would proceed to look up or otherwise figure out the answer. Try to think about associations and the fundamentals underlying the question, then show your knowledge of those associations and fundamentals. In some cases, you might outline a study to address the problem. Remember that in a good exam, you will be asked questions that committee members know are outside your range of knowledge.

- Keep your cool, remain calm, think before you speak.

- Defend your work but don't become defensive.

By the time you have successfully completed all course work, passed the qualifying exams, and finished the research you should be well prepared for this final hurdle. No reasonable committee would want to flunk you at this point—it could reflect as badly on them as on you—nevertheless they could send you back to do some revisions. By issuing a passing grade, the committee certifies to society that you are competent as a beginning professional, and in the case of the doctoral degree, they further certify that you are ready to be accepted as a peer.

CHAPTER 2

On Definitions and Dualities

What's This Chapter About?

Definitions for direction

Philosophy of research

Tools to inspire and direct

What's in This Chapter?

What Is a Thesis and What Is Research?

A Brief History of Scientific Research Philosophy

Definition Dualities: Thought Tools for Research and Problem Solving

Definition Dualities: First Principles

  Rationalism vs. Empiricism
  Rationalism vs. Rationalization
  Deduction vs. Induction
  A Diversion to the Game of Rationalism-Deduction
      vs. Empiricism-Induction
  Daydreaming vs. Systematic Inquiry
  Science vs. Advocacy

Definition Dualities: Design

  Basic vs. Applied Research
  Fact vs. Assumption
  Theoretical vs. Empirical Research
  Analysis vs. Synthesis
  Reductionism vs. Holism
  Analytical vs. Predictive Studies
  Theoretical vs. Operational Definitions

Definition Dualities: Outcome

  Conclusion vs. Inference
  Validity vs. Soundness
  Correlation vs. Cause-Effect
  Results vs. Interpretations

A Triad: The Testable, the Trusted, and the True

  Hypothesis
  Theory
  Law
  And Then There's the Paradigm

Speaking of physical chemistry—it's easy if you know what you're talking about.

From the first-day lecture by a chemistry professor who shall go unnamed

I n any advanced intellectual endeavor, including scientific research, you need to know not only what you're talking about, you also need to know how to *think* about what you're talking about. For most of us, critical and productive thinking doesn't come naturally—it requires training and practice. One of the best ways to acquire that training and practice is by conducting original research for a graduate thesis.

This chapter presents concepts and skills you should learn to start appropriately, proceed effectively, and finish competently. It begins with a definition of thesis and of research, each of which includes criteria that your work should meet. Next it summarizes a bit of history behind these definitions. The history introduces a few people and ideas from the nucleus of scientific research philosophy. Finally, it makes practical sense of the philosophy by putting forth paired options for guiding your thinking and making choices. We'll call these pairs definition dualities. Whereas the thesis and research definitions broadly outline where you will need to go, the dualities offer you choices in how to get there. Consider the dualities carefully—they are intellectual tools you will need for the job.

What Is a Thesis and What Is Research?

"Thesis" is defined in most university catalogs or other academic governing documents. One good definition is that written in Title 5 of the California Code of Regulations (n.d.), which sets basic requirements for public universities in the state. Pay careful attention to the definition. It identifies the components and criteria of a graduate thesis:

> A thesis is the written product of a systematic study of a significant
> problem. It identifies the problem, states the major assumptions,
> explains the significance of the undertaking, sets forth the sources
> for and methods of gathering information, analyzes the data, and
> offers a conclusion or recommendation. The finished product evi-
> dences originality, critical and independent thinking, appropriate
> organization and format, and thorough documentation. Normally, an
> oral defense of the thesis is required.

Although the word "research" is not mentioned, it is implied in the above definition. But what is research? Does it require experimentation or some other form of direct investigation, or can it be carried out entirely in libraries? The answers vary widely among professions and even among schools of thought within a profession. We will consider research to be "critical and exhaustive investigation or experimentation having for its aim the discovery of new facts and their correct interpretation" (*Webster's Third New International Dictionary of the English Language, Unabridged*, 1971). Several points of this definition should be noted:

- Research might or might not involve experimentation. This option allows for different approaches, including theoretical studies designed to formulate models or to characterize systems without collecting data or directly manipulating physical variables. It also allows for characterization studies in which data are collected and analyzed but variables are not manipulated.

- Research need not be successful in reaching a desired end, rather it *aims* at a desired end. This aspect of the definition emphasizes process over product. Whether the results are positive, negative, or inconclusive, the effort, if properly conducted, still can be research.

- The definition reads "the discovery of *new* facts," which seemingly would exclude purely library work. But does it? Perhaps not, if the library investigation produces a fresh analysis and integrates previously unconnected facts into new discoveries. Still, simply writing a literature summary is not scientific research, even though the process might be called library research. (See "Preparing a Literature Review," in Chapter 4.)

- Facts require *correct interpretation*. This criterion takes research beyond the mere collection and presentation of data. It establishes the need for a thoughtful design and an analytical method for evaluating and interpreting the data. Note that correct interpretation does not necessarily mean free from error, rather it means that interpretations be done properly and meet standards of the discipline. Making interpretations, whether they are qualitative or quantitative, can be one of the most challenging and controversial obligations the researcher faces. Quantitative approaches often employ statistical analyses, which can help estimate the reliability of numerical results.

A Brief History of Scientific Research Philosophy

There are more things in heaven and earth,
Horatio, than are dreamt of in your philosophy.

William Shakespeare, *Hamlet*, Act I, Scene V

Shakespeare's words could not have rung more true in his time, a time near the end of the Renaissance that witnessed gradual but powerful changes in European philosophy and its impact on scientific advancement. The changes, which spanned many decades on both sides of Shakespeare's life, were exemplified in part by a rising secularism and the clash of emerging scientific thought with long-standing religious authority.

For about a 250-year period, between the sixteenth and nineteenth centuries, the inquisitional tribunals of the Roman Catholic Church invoked centuries-old Roman legal procedures to protect their religious beliefs from what the inquisitors perceived as heretical attacks. Anyone found guilty of espousing opinions that contradicted religious teaching could either recant and have his or her life spared or suffer under an auto-da-fé, a ceremoniously elaborate public sentencing and execution by burning at the stake. Although the inquisitions were centered in Mediterranean Europe, their influence extended into the northern countries, including Germany and Britain. Ultimately the effects reached North America, albeit under a different authority and ideology (Peters, 1988).

Heresy could take many forms, including speculations and direct observations in natural science that went against biblical accounts or religious dogma. A good example lies in the discordant explanations of comets. To devout theologians, who took the Bible literally but who also applied their own peculiar logic to Scripture and worldly events, comets were omens that originated from somewhere this side of the moon and passed threateningly close to Earth. They were seen as "heralds of Heaven's wrath"—signs of God's displeasure with human sins and warnings that punishment soon would follow (White, 1898). All souls had better repent, or watch out.

In contrast, a few diligent observers noted that comets originated from far beyond the moon and that they followed regular, harmless paths across the sky. Although these nascent empiricists inferred that comets were natural bodies, most of them were cautious in presenting their ideas. Anyone having a strong preservation instinct was sure to strike a careful compromise between science and theology. Some were not so guarded and paid dearly for their transgressions.

So imagine yourself a modern-day graduate student in astronomy studying the nature and orbits of comets. For your degree, you must conduct original research that discovers new facts and interprets them correctly. Of course you must prepare a thesis that meets the criteria given in the definition above.

Through the magic of a time machine, you get the opportunity to make first-hand measurements of a comet that passed by Earth in 1572. So you load up your equipment and head back in time. Soon after arrival, you learn that your "older" colleagues find your notions of research to be as strange as your high-tech equipment. They shun you, fearful that any association would bring the wrath of the Reverence down on them.

Word of your arrival spreads, and before long, an inquisitor pounds at your door. He demands to see your guiding principles—that is, your definitions of thesis and research. You are in big trouble. Suddenly, those definitions no longer look routine and innocuous; rather, they are a threat to the establishment. Your visitor never dreamt of them in his philosophy, and you realize that they are indeed revolutionary.

This little scenario illustrates how scientific approaches that today we take for granted were restrained for centuries by the long-standing discord between spiritualism and secularism. It also demonstrates that the controlling beliefs and philosophy throughout much of Continental Europe before the 18th century would not have sanctioned the type of research we do today. Undoubtedly, those beliefs would have discouraged the very thought processes that guide today's scientists. These processes, which will be represented by the definition dualities, constitute some of our most useful thinking techniques. To help you grasp and appreciate these intellectual tools, we will review a bit more of their philosophical background.

The most fundamental concepts stem from aspects of Western philosophy, namely the nature, sources, and limits of knowledge (epistemology), the nature of reality (metaphysics), and the principles of reasoning (logic) (Woodhouse, 2006). You will enter the realm of these philosophic divisions many times during your graduate studies, whether you are aware of it or not. For example, whenever you must determine the reliability of your knowledge, including your research findings, you will employ epistemology; when you try to distinguish and make connections between reality and abstraction, as in interpreting results, you will be dealing with metaphysics; and when you must ensure that your reasoning is sound, you will use logic. Most of the time, these divisions probably will blend together so that you won't even recognize them. That's normal; still, you should develop awareness of their relevance.

Open inquiry, including research planning and hypothesis testing required of most modern-day graduate students, once was severely punished.

Two of the earliest and most debated philosophical dualities, which originated with Plato and Aristotle, are rationalism vs. empiricism and deduction vs. induction. These terms will be discussed in the next section, but for now, brief definitions will suffice. Whereas *rationalism* is the belief that knowledge stems from intuition and reasoning, *empiricism* is the belief that knowledge originates from experience. Rationalism then relies on *deduction,* which is the process of using reasoning from beginning principles to explain specific observations. Conversely, empiricism utilizes *induction,* or the process of formulating a general principle from specific evidence. In logic, an outcome that is rationally derived has greater certainty than one that is empirically inferred.

Although sown by ancient Greeks, the seeds of these fundamentals lay virtually dormant until the work of Renaissance scholars stimulated their germination nearly a millennium and a half later. Still more generations passed before their development finally took firm root in the early 17th century, when European philosophers began debating them in earnest and the first modern scientists challenged philosophic and theologic supremacy.

Prior to the 17th century, during medieval times and well into the Renaissance, experimentation and inductive logic were not viewed as scholarly activities; rather, deductive reasoning from religious faith and related metaphysical principles dominated advanced thinking. These parochial trends had begun to take hold after collapse of the Roman Empire in the 5th century, following which, populations dispersed and soon lost their intellectual amenities and their cultural identity. Gradually, they regrouped into scattered castles and monasteries, which became the new centers of civilization.

Scholarship became confined almost exclusively to monasteries. Monks, who devoted their lives to studying scripture, transcribing manuscripts, and indoctrinating novices, had little opportunity for worldly experience. Of course they became well practiced at the empirically driven activities of growing food and making clothing, but for what they felt they needed to know most, they relied on reading, dialogue, and abstract thinking. Rationalism and deduction ruled their intellectual domain.

By the 13th century, life was being reorganized into and around cities, and scholarship was being taken over by newly established universities. The universities adopted the monastic learning and teaching style—empiricism and induction had no place in the curricula.

While life in the monasteries was devoted to devotion, life in the castles centered on day-to-day challenges of eating, raising children, suffering disease, and supporting the lord's economic system. Castle residents faced practical problems that could not be solved by grand thinking of cloistered monks. They relied on trial and error to guide their efforts in agriculture, trades, navigation, and combat. With time, entire sophisticated industries rose from their efforts. Technological development born of trial and error was fast outpacing social development, which had no comparable methodology (Zilsel, 1941). The trend continues today.

Although often creative and successful, the trials of early empiricists could not be considered science, or even research. They consisted largely of haphazard observations made by uneducated people, who were unschooled in controlled experiments and systematic inquiry. Still, in many respects their efforts ran far ahead of those of the intelligentsia.

As empiricism developed, rationalism and empiricism eventually came to be seen as disparate views, not only about the source of knowledge, but perhaps more importantly, about truth and acceptable methods of investigation. Experiments, which required hands-on examination, were seen as lower-class activities and beneath the dignity of learned society. Surely, in the minds of the elite, such activities could never elicit truth. Accordingly, religious and political leaders scorned empiricism, especially if evidence or the ideas it generated ran counter to the prevailing dogma and suggested a questioning of religious or political authority.

The two doctrines continued to evolve over several centuries of discourse in Western philosophy and science. The truth or falsity, merits and demerits, piety or blasphemy of the processes and products of each were debated, sometimes vehemently, on both sides of the English Channel and judged by the Inquisition. (In the 19th century, the inquisitions became subjects for powerful works of art, including Francisco Goya's paintings and Giuseppe Verdi's opera *Don Carlos*, which was based on an earlier literary work by Friedrich Schiller.)

The seminal arguments from the likes of the mathematical French rationalist René Descartes (1596–1650) and the less mathematical British empiricist Francis Bacon (1561–1626), combined with the great scientific achievements and sacrifices of Galileo (1564–1642) to initiate what eventually emerged as modern concepts of science, scientific methods, and research.

Of course, many more brilliant philosophers contributed to the development of these ideas. Perhaps most notable among 17th- and 18th-century deductive-rationalists who built on Descartes' work were Spinoza (1632–1677) and Leibniz (1646–1716). Prominent inductive-empiricists included Locke (1632–1704), Berkeley (1685–1753), and Hume (1711–1776). Although each of these philosophers has been identified as either a rationalist or an empiricist, none seems to have subscribed purely to either view. More correctly, each favored one theory over the other, while acknowledging value in the opposing viewpoint (Russell, 1945; Cottingham, 1988; Woolhouse, 1988).

Scientists of the day found themselves in the middle of the philosophic disputes. One of the most famous was the Italian Galileo Galilei. Early in the 17th century, Galileo (who was born in 1564, the same year as William Shakespeare) took a giant step toward modern science when he broke intellectual restrictions by conducting experiments in physics and mechanics and by interpreting planetary movements based on mathematics and direct telescopic observations. Galileo's work confirmed the Copernican theory that Earth revolves around the Sun. Unfortunately for him, his empiricism and reasoning went against the Church-supported Ptolemaic–Aristotelian view of an Earth-centered universe.

For his efforts, and because he had antagonized the clergy and violated a papal injunction against such experiments and pronouncements, Galileo eventually was arrested on heresy charges. He stood before an inquisition in Rome, where he was found guilty and threatened with torture lest he recant his astronomical findings. He also was required to give up the bulk of his research and disavow his ideas for advancing knowledge. Galileo complied and was sentenced to house arrest. He could continue limited studies and writing in mathematics and physics, but he also suffered humiliation and misery until he died in 1642 (White, 1898; Peters, 1988).

Although Galileo's conviction essentially ended his extraordinary scientific career, his efforts and accomplishments nonetheless fostered a shift from locked-in, church-enforced rationalism and associated deductive reasoning to innovative inductive approaches based on free imagination, systematic observation, and measurement. His remarkable creativity, courage, and determination influenced all of science. Galileo's methods and admirable characteristics were shared to some degree by his eminent contemporaries, including Johannes Kepler (1571–1630) and Robert Boyle (1627–1691).

In the late 17th century, mere decades after Galileo's death, Isaac Newton (1642–1727) carried science another giant step forward when he formalized rigorous experimental science and more strongly linked inductive and deductive approaches into what has become known as the scientific method. (A misnomer—any good scientist knows that science does not progress by a single method; still, we often use the term for convenience.)

These scientists made enormous contributions to scientific progress and human understanding. They were imaginative and resourceful, but they also learned to channel their efforts through trained and organized work. They coupled creativity with pragmatism and transformed fantasy into reality when they revolutionized intellectualism by infusing it with systematic inquiry.

The revolution started by the 17th-century scientists and philosophers ebbed and flowed through several generations and into the Enlightenment period, beginning in the mid-18th century. At this time, the influential German philosopher Immanuel Kant (1724–1804) vitalized and formalized the debate by clearly distinguishing rationalism and empiricism while concurrently advocating their interdependence in human thought (Kant, [1787] 1900). Kant argued in part that experience can stimulate ideas, but rational thought completes their development. Alternatively, intuition (i.e., rationalism) can generate ideas, empirical evidence can support or refute their validity and guide their further development, and rationality can identify their applications and interpret their value.

Gradually, through the 19th and early 20th centuries, empiricism-induction and rationalism-deduction gained increasing—but by no means universal— acceptance as equally valid and useful approaches to building scientific knowledge. By mid-20th century, multitudes of researchers had adopted, refined, and expanded the advances of Galileo, Newton, and other notable innovators. Modern research scientists routinely employ a complement of empirical inputs (i.e., evidence gathering) and rational efforts (i.e., reasoning) to generate ideas, choose actions, collect facts, interpret relationships, and reach defensible decisions.

Indeed, our definitions of thesis and research presented above embody these innovations. These definitions, as well as our freedom to study and expand our knowledge beyond artificially imposed limits, would not have been possible without the insights, determination, courage, and sacrifices of the individuals who dared to oppose the contemporaneous status quo. More of the same is needed for the future.

Despite earlier advances, arguments over rationalism and empiricism as valid sources of knowledge and over deduction and induction as distinct and legitimate scientific approaches continued through the 19th and 20th centuries. In the late 1800s, intellectual disputes between William Whewell ([1860] 1971) and John Stuart Mill ([1843] 1874) dominated the discourse. Nearly a century later, podiums were occupied by a host of philosophers and scientists, including Karl Popper (1959, 1962), Rudolf Carnap (1962), Carl Hempel (1966), Peter Medawar (1967, 1969), Paul Feyerabend (1975), and Thomas Kuhn (1996). Writers in this latter group were most active during the three decades following World War II, a time that saw a surge in science-philosophy debates coupled with extraordinary scientific achievements, including the discovery of DNA's molecular structure, the dawn of the space age, and the emergence of computer science.

Serious differences stemming from rationalism and empiricism as foundations of values and beliefs continue to extend beyond scientific philosophy to theology, politics, and a broad spectrum of social discourse. Although these types of differences are outside the realm of scientific research (and beyond the scope of this book), they nonetheless continue to influence research and scientific education.

The concepts presented here and the people who developed them can be studied in depth at universities throughout the world. The writings of the great contributors are readily available, and additional volumes have been written about each and all (e.g., Russell, 1945; Whewell, [1847] 1967; Cottingham, 1988; Woolhouse, 1988; Zalta, 2004). The main purposes here have been to outline the history of some underlying principles in science and help you appreciate intellectual connections between science and philosophy. In the next section, you will learn how to make practical use of these and related fundamentals.

Definition Dualities: Thought Tools for Research and Problem Solving

Neither the bare hand nor the unaided intellect has much power; the work is done by tools and assistance, and the intellect needs them as much as the hand.

Francis Bacon, *The New Organon*

For most people, the word "tools" probably suggests tangible devices designed to accomplish physical tasks. But in the problem-solving fields, including scientific research, other types of tools are more fundamentally necessary than any hardware. These are the intangible, abstract, and all-too-elusive mental devices that can help us direct, clarify, and vitalize our thinking: these are our "thought tools."

Some of the most effective aids for constructive thinking—ones that help us generate new ideas, set a proper course, and render impartial judgment—are the type that prompt us to recognize and evaluate opposing alternatives. For this reason, the thought tools in this chapter are presented as definition dualities. That is, they are paired as contrasting options for thinking about, designing, and evaluating the various parts of your study.

Admittedly, most established researchers probably pay little heed to these dualities—at least consciously and purposefully. More likely, they go about their work unaware of the thought processes they employ (Medawar, 1969; Cohen and Medley, 2005). Most learned their craft by practicing under a mentor in graduate school followed by personal experience. They became keenly aware of the technical aspects of research methodology, but remained largely indifferent to their intellectual methodology. Although they might be creative, competent, and productive, many researchers could do even better if they analyzed their avenues of thought, rather than leaving that job to outside philosophers.

Many of the concepts presented have been intensively discussed in numerous books and articles, as well as in a multitude of internet sites (and, unfortunately, too often with insufferable verbosity and convolution, which turns off even the best of scientists [Feynman, 1999]). Probably the most celebrated of the concepts are rationalism, empiricism, deduction, induction, reductionism, holism, validity, soundness, and paradigm. A main goal here is not to present all the types, gradations, and fine distinctions of each concept; rather, it is to keep the presentations brief, readable, and practical. If you choose to dig deeper, as you should, you will find numerous variations and views of each. The literature on these subjects can be confusing and contradictory but nonetheless scattered with real gems of insight and wisdom. As your studies of philosophy progress, keep your research moving—don't allow conflicting viewpoints in the *philosophy* of science to hinder your *doing* of science.

The dualities are divided into three groups: (i) those considered common or first principles, (ii) those that tend to fall on the design side of a study, and (iii) those that pertain more to the outcome side. Although each duality is assigned to its primary group, some pairs can overlap among groups. For example, the important first principles rationalism vs. empiricism and deduction vs. induction can apply to both the design and outcome sides of your study. Fact vs. assumption is presented on the design side but also applies to the outcome side, and correlation vs. cause-effect pertains to outcome but should be anticipated when designing a study.

In considering the options within a duality, you might select one exclusively over the other, such as in identifying your research as either basic or applied; but in most cases you should alter your focus deliberately from one to the other. You might, for example, shift back and forth between theoretical and empirical aspects of your study. Alternating your thinking and approach in these ways will help keep you fresh and infuse a system of checks and balances into your project. In some cases, it might even help you clarify the association between abstraction and reality.

As you work with the dualities, you will find some to be straightforward and easily applied, while others will be more elusive, requiring deliberation and experience to grasp fully and employ effectively. Whether the options seem obvious or obscure, you should consider each pair carefully. They can help you strengthen and expand your thinking repertoire as well as help you recognize and choose alternate paths when you get stuck.

Definition Dualities: First Principles

This section presents the most basic dualities—ones that are universal and should be carefully considered throughout the course of any study.

Rationalism vs. Empiricism

> *The origin of discoveries is beyond the reach of reason.*
>
> W.I.B. Beveridge, *The Art of Scientific Investigation*

What do our minds tell us through reason? What do our senses tell us through experience? How do we come to know things anyway? Scholars have pondered these questions since the dawn of philosophical thought. Centuries ago, philosophers answered with two radically different theories: *rationalism* and *empiricism.*

From historical and philosophical perspectives, these terms embody contrasting beliefs about the origin of knowledge. Although the ideas are not mutually exclusive, philosophers normally set them apart in efforts to organize what we have learned about how we learn (See the previous section, "Brief History of Scientific Research Philosophy").

Rationalism is the belief that knowledge derives from reason and logic, rather than from experience. A strict rationalist would assert that what we know—facts, understandings, truths—comes to us intellectually. We learn by rational thinking from innate, a priori (i.e., without previous knowledge or experience) conceptions. Worldly details and phenomena that our senses detect can be figured out or explained by deduction from these conceptions. Simply put, everything of importance already is in our brains, we just have to pull it out and put it to use solving problems.

Rationalists also maintain that knowledge gained in this way is logically superior to anything that might be learned through sensory experience. They argue that although experience might promote awareness and conditioning, it cannot generate true knowledge. Our senses are too erratic and unreliable, and we can never gather enough data to ensure certainty.

Empiricism, in contrast, is the view that knowledge derives a posteriori; that is, it builds from sensory experiences. A strict empiricist would argue that, except perhaps for mathematics, we learn solely from experience—our senses accumulate information that our minds interpret and transform into knowledge. Understanding is built up inductively from collected facts. In short, whereas rationalists learn by thinking, empiricists learn by doing.

Strictly defined, *rationalism* and *empiricism* refer to beliefs or theories that are favored one or the other by somewhat opposing groups. For centuries, these approaches have set the course of individual endeavors and, in a larger sense, the course of empires. (Consider, for example, that the great rival economic systems of the 20th century, capitalism and communism, arose from empiricism and rationalism, respectively.) We will stretch the definitions slightly and allow the terms also to suggest *processes* of generating knowledge. Taking this small linguistic liberty will allow us to use the terms not as disparate ideas in a philosophical dispute but rather as contrasting, yet congruous, cognitive choices. In this way, the terms can identify alternative thinking styles or "thought tools."

As thinking styles, rationalism and empiricism can complement each other to produce knowledge. In this sense, creating knowledge can be analogous to transferring energy in basic physics. Any energy transfer involves an intensive and an extensive component. For example, the work done in pushing an object is the product of the force applied (intensive) times the distance pushed (extensive). When we apply this analogy epistemologically, rationalism becomes the intensive component and empiricism becomes the extensive component. Multiplying the two together produces knowledge (Fig. 2-1).

$$\text{Work} = \underset{\text{(intensive)}}{\text{Force}} \times \underset{\text{(extensive)}}{\text{Distance}}$$

$$\text{Knowledge} = \underset{\text{(intensive)}}{\text{Rationalism}} \times \underset{\text{(extensive)}}{\text{Empiricism}}$$

Figure 2-1. Knowledge creation, like an energy transfer, comprises intensive and extensive components.

Of course, the intensive and extensive components need not be equal. Just as we could increase the work done on an object by pushing harder or by pushing farther, we can increase our knowledge through more intensive thinking or more extensive experience.

Empiricism and rationalism complement each other in ways that the simple work equation does not reveal. Empiricism can stimulate alternate points of view, show the value of evidence and replication in making judgments, and provide quantification and precision needed to transform thoughts into physical reality. Rationalism, in turn, can strengthen empirical efforts by adding diversity, organization, meaning, and certainty. Acting together, each enhances the other to create a synergism, in which the whole is greater than the sum of its parts.

With conscious effort and practice, you can learn to shift your learning mode between empiricism and rationalism (i.e., between evidence gathering and reasoning). In so doing, you can reorient your point of view as well as your thought process to best deal with the problem at hand. Recall for example that thesis and research entail the collection, analysis, and interpretation of information. The dissimilarity among these activities obliges you periodically to reorient your thinking and learning modes. This is when the duality becomes a thought tool.

You might consider the two learning styles as end-members of a continuum, with a slider between them (Fig. 2-2). Here's how the device works. With your

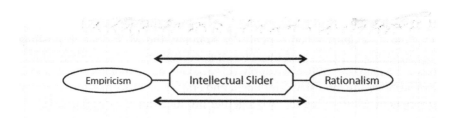

Figure 2-2. Use an intellectual slider to shuttle between dualities.

intellect riding on the slider, direct it to and fro as needed to adjust your course. When evidence is lacking, slide to the empiricism side; when more interpretation and innovation are called for, slide to the rationalism side. When you're stuck and don't know what you need, jiggle it a little each way until something gives. Under continued use the slider should not lock up at either end of the track.

Now you might ask, But what if it does lock up? How can that happen? How will I know? What are the consequences? Does it really matter? After all, it's only an abstract idea!

Sure, it's an abstract idea, but remember—all thought tools are abstract. If you want to learn science, conduct research, and most importantly expand your thinking skills, you must acquire the tools and apply them competently. So it does matter that you exercise the slider regularly and purposefully. If you don't, you'll end up in a rut.

Bear in mind that the purpose is not to maintain a steady-state or static balance by hovering the slider near midpoint. Before long, that effort would turn dull and uncreative (although it could be reasonably productive for a time). Instead, you should strive for a dynamic equilibrium over the long term by applying mental effort to power it to and fro between the extremes.

Probably each of us used a similar device early in childhood as we responded and adapted to sensory inputs and experiences while acquiring ability to reason. An important difference between children and adults is that children in their formative years don't need to power the slider between experience and reason. It oscillates naturally as the child develops mechanisms for coping with life. Once developed, the mechanisms gradually can lock in and become functional habits for self-protection (albeit not always constructive ones in the long run). And the slider, now neglected, ceases to operate.

So in adulthood, whether your early conditioning prepared you to greet the world openly or with caution, the slider no longer is self-propelled—you must move it willfully. If you neglect it, it can stagnate. But it won't necessarily lie where you left it. Most likely, it will drift spontaneously to the right, where it can lodge and atrophy. Eventually, without a valid empirical counterbalance, reasoning can become overused, under challenged, and unsupported as it gradually degrades to *rationalization*, a type of false rationalism designed to justify one's thoughts and actions.

Rationalism vs. Rationalization

Rationalism implies a genuine attempt to develop communal knowledge through dispassionate reasoning. Being intellectually driven, it can be honest, insightful, and creative. In contrast, *rationalization*—or false rationalism—connotes a defensive mechanism that perpetuates individual or group prejudice. It is more emotionally driven and too often is self-satisfying, self-serving, and incorrect.

While evidence may be considered in rationalizations, it is not evaluated objectively. Instead it is filtered and weighed selectively to justify a position or to support a belief (Evans and Feeney, 2004). The intellectual slider when driven by this motive becomes corrupted. It degrades to an inferior instrument—one of pretentious and empty justification.

Rationalizations, then, involve premises that are wishful and believable, but not necessarily defensible. They can be proposed with or without valid evidence. Accordingly, they suggest bias and unfounded assertion to support preconceived notions or to gain advantage. Because rationalizations are plausible and self-satisfying, they can charm the intellect and override detached reasoning. They can become persuasive and habit-forming, regardless of whether they are sound or not. In this debased state, so-called rational efforts are not genuinely rational. They contribute little to real knowledge, largely because they confuse knowledge with belief, rendering the two virtually indistinguishable—and once accepted are extremely difficult to change.

Does all this mean that rationalizations are inherently "bad"? Not necessarily. Even top thinkers probably rationalize something nearly every day as they seek relief from life's pressures. While trivial rationalizations can help us cope with daily challenges, they have little effect on the development of knowledge and the course of science. A scientist should recognize the difference between a substantive and a trivial rationalization, and more importantly should recognize when rational integrity is being compromised.

Deduction vs. Induction

By conventional definitions, *deduction* and *induction* are modes of reasoning or speculation used to formulate philosophical arguments and to draw

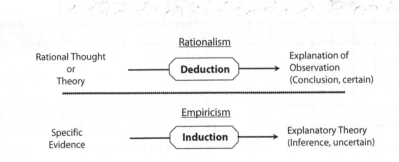

Figure 2-3. Contrast between deduction and induction.

conclusions or inferences. They differ in "direction" of reasoning and in certainty of the outcome (Fig. 2-3). Deductive arguments, which emerge from rationalism, use general principles or theories to explain specific observations or phenomena. They produce resolutions that are guaranteed true, provided their premises are true. In contrast, inductive arguments, which are generated empirically, amass specific evidence to propose general principles or theories. They can lead to decisions that might have a good probability of truth but that leave some doubt because they are not perfectly verifiable (Kotarbinska, 1962; Hempel, 1966; Stenning and Monaghan, 2004; Oaksford and Chater, 2007).

One way of presenting deductive arguments is through syllogisms, which are logical structures consisting of a major premise, a minor premise, and a conclusion. Although these structures are of little value in most scientific research, they are fundamental to logic and can help us better understand deductive and inductive thought trains. Syllogisms often pertain only to deductive arguments; nonetheless, we will use the format to help distinguish deduction from induction. (See Fig. 2-4 for an example of each.)

In the deductive example (Fig. 2-4), the conclusion that Galileo maintains at least a B average is logically certain provided the major and minor premises are true. But what if Galileo's grades dropped and he was allowed to remain in the program on probationary status? In this case, the conclusion would be false. Doubt would be cast on the major premise, and it would have to be modified; nonetheless, the argument's structure is still deductive.

In the inductive example, the inference that typical graduate students maintain at least a B average is less definite than the deductive conclusion about Galileo for at least three reasons: (i) only a small sample of graduate students was evaluated; (ii) not all graduate students are necessarily typical in the same ways; and (iii) we cannot guarantee that conditions and events of the past will continue universally unchanged into the future. The first limitation

Deduction	Induction
Major premise: All graduate students maintain at least a B average.	Major premise: Galileo and Marie are typical graduate students.
Minor premise: Galileo is a graduate student.	Minor premise: Galileo and Marie each maintain at least a B average.
Conclusion*: Galileo maintains at least a B average.	Inference*: Typical graduate students maintain at least a B average.

*See also "Conclusion vs. Inference," later in this chapter.

Figure 2-4. Syllogistic examples of deductive and inductive reasoning.

could be tempered by increasing the sample size, but some degree of logical uncertainty would remain unless the entire population of graduate students was sampled. The second limitation could be addressed by defining the terms "graduate student" and "typical," and ensuring that all students in the population (not merely in the sample) perfectly fit the criteria. The third limitation could be resolved by freezing time—you can guess the chances of that.

As applied to scientific research, deduction and induction suggest different basic approaches for expanding knowledge and solving problems. Deductive science, which is more *intensive* (refer back to "Rationalism vs. Empiricism"), proceeds from the general to the specific. By this approach, the scientist employs known principles to help explain specific observations or other data, or to predict possible results. If a known principle (i.e., the premise) is true and the project proceeds properly and without error, the conclusion (i.e., the explanation or prediction) also is true.

In the absence of an established principle, the scientist still can proceed deductively by proposing a hypothesis to explain the phenomenon of interest. By this method, the hypothesis (i.e., a speculation or proposal whose certainty is unknown) becomes the major premise. It then is tested by carefully controlled experiments that incorporate minor premises as well as background knowledge, including reasonable assumptions. If the experimental results support the hypothesis, the results may be said to corroborate it, but not to verify or prove it. Although the supported hypothesis might be compelling and useful, it still is logically uncertain and subject to further testing. (See also the section "Hypothesis," near the end of this chapter.)

Deduction vs. induction: the case of the undecided dogs. Given what we know about dogs, rational deduction assures us that Hamlet and his pals will select one of the two choices before them. But which one? We might observe a few of the crowd and develop a predictive model based on their actions. The modeling effort entails empirical induction, and its prediction remains eternally uncertain.

Conversely, if experimental results falsify the hypothesis, the hypothesis is shown to be untrue. No further testing is needed, except perhaps to ensure that no mistakes were made in conducting the experiments.

Because falsification always is more certain than verification, hypotheses should be stated in such a way that falsifying them is at least possible. Although a hypothesis may be appealing and robust, it also should be liable to being proved wrong; otherwise it is scientifically invalid. In addition, testing should be rigorous—that is, the researcher should strive to disprove the hypothesis, not to prove it. This technique, sometimes referred to as the hypothetico-deductive method, is one of the most powerful and commonly used tools in science; it is at the heart of scientific research. (For more in-depth discussions and historical perspectives see Kant, 1900; Popper, 1959, 1962; and Medawar, 1967, 1969.)

Once falsified, the hypothesis may be revised (i.e., repaired or replaced by a new, improved version) and retested, multiple times if necessary. This revise-retest process, which is a sort of reciprocation between outputs and inputs, constitutes a negative feedback loop—a mechanism through which

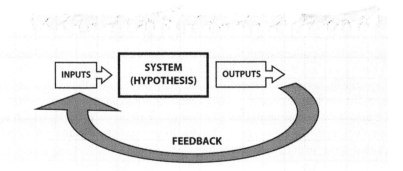

Figure 2-5. A hypothesis is analogous to a system that can be transformed by interchanges with its environment. The environment provides inputs (e.g., data amassed by the researcher) to the system, and the system releases outputs (more data) to the environment. If an output somehow reinforces an input effect, the process is a positive feedback. Alternatively, if the output opposes or counteracts the input effect, the process is a negative feedback. Positive feedback loops tend to be self-destructive; whereas negative loops, which generate checks and balances, are more adaptive. In hypothesis testing, a scientist should favor negative feedback.

outputs (i.e., testing results) are fed back as inputs into the hypothesis, allowing it to be adjusted, strengthened, and retested (Fig. 2-5). The feedback loop remains negative as long as the new data produce a result that is somehow opposed to that produced by earlier data—in other words, some aspect of the hypothesis continues to be falsified. This technique, which also provides a system of checks and balances in research, is a powerful engine for moving science forward.

Despite its advantages, the revise-retest process can be misused. While repeated tweaking might produce a more resilient hypothesis, it eventually can defeat testing if the objective, deductive approach is inadvertently replaced by a more subjective, inductive approach. Accordingly, the hypothesis then is made to fit the data, and a point can be reached at which analysis converts to synthesis and rigorous tests of soundness give way to fabrication.

If the hypothesis is made to fit the data, the feedback loop converts from negative to positive because new data amplify, rather than oppose, earlier inputs. In many cases, this conversion can be enlightening and appropriate. Indeed, much of research science proceeds in this manner—scientists call it model building. Still, the models should be tested deductively. As a researcher, you should endeavor to be aware of the processes you use, and to recognize when you cross over from one style to another.

In contrast to deductive approaches, inductive science proceeds in the opposite direction and is more *extensive* (refer back to "Rationalism vs. Empiricism"). The scientist gathers empirical evidence through direct observations or other means and then assembles the information to propose a new hypothesis. The hypothesis, which is a type of model, commonly predicts cause-effect relationships or future conditions or events. As noted earlier, the hypothesis must be tested by experimentation or some other rigorous process, which usually involves more data collection. The results are statistically analyzed and interpreted with respect to the hypothesis.

Accordingly, scientific hypotheses and theories can be formulated and supported by induction, but when these are tested or later are used as premises for new conclusions, the reasoning is deductive. Overall, this procedure of observation, hypothesis formulation, experimentation, analysis, and interpretation commonly is known as the scientific method (Gauch, 2003). To most modern scientists, this method is neither solely inductive nor solely deductive; rather scientists see the two approaches as necessary and complementary, even though deduction dominates logically.

Inductively generated principles are inherently uncertain because scientists cannot gather an infinite amount of data, which presumably would be needed to support every possible instance of the principle. This logical limitation notwithstanding, inductive approaches to problem solving can be fruitful and sufficiently reliable to challenge and shift long-accepted paradigms (i.e., seemingly fixed or tradition-bound scientific models that set direction within a discipline) (Kuhn, 1996). A shift in paradigm can alter the normal course of an entire field of study. Accordingly, induction has stimulated some of the most significant advances in science, including Newton's Laws and Boyle's Law.

Despite these long-established and widely accepted meanings of deduction and induction, the use and logical validity of induction has been a matter of disagreement among philosophers of science (e.g., see Hume, [1739] 1967; Popper, 1959; Carnap, 1962; Kotarbinska, 1962; Pears, 1990; Kuhn, 1996; Zalta, 2004). Some highly regarded philosophers and scientists have argued fervently that the role of induction in advancing science is purely a myth (Popper, 1962) or at best inadequate (Medawar, 1969).

They reason that because a scientist always has a frame of reference for making observations, the observations cannot be made without prejudice. The scientist has a purpose, which arises from conjecture about the nature of whatever it is he or she wishes to observe. Thus, the act of observing (as opposed to random gazing) implies the presence of a hypothesis, albeit one that might be only vaguely or even subconsciously conceived. This being the case, the scientist always proceeds deductively, no matter how ill defined the process.

Critics also recognize that making observations (i.e., collecting data) prompts the scientist to define the hypothesis more clearly. Additional data are collected to augment earlier data. This technique not only *establishes* a result, but it also leads to *justifying* the result (in effect creating a positive feedback loop); thus the separate processes of discovering facts and justifying results run together. *In science, these processes should be discrete and clearly distinguished.*

Other scientists contend that, despite its inherent uncertainty, induction is key to advancing scientific knowledge. They strongly uphold the value of inductive generalizations and often describe scientific method in inductive terms (even if some do not use the word "induction" directly). (Hempel (1966) gives a nicely balanced account. See also Lakatos, 1978; Tweney et al., 1981; Kuhn et al., 1988; Salmon, 1998; Cohen and Medley, 2005; and Johnson-Laird, 2006.)

Adequate data and proper statistical analysis can establish a logical relation between an inductive inference and its evidence. If the outcome meets a predetermined standard and satisfies the appropriate statistical rules, the inference will merit confidence at a specified probability. Statistical methods also allow competing hypotheses to be mathematically compared. Thus the arguments are made that a hypothesis can be defended, and that induction stands as a valid and useful means of advancing knowledge.

So what does this philosophical dispute mean to you and your work? As a graduate student in either the natural sciences or social sciences, you should have some awareness of scientific philosophy, and you should be able to appreciate opposing points of view. You should recognize when you are violating logical precepts (i.e., being illogical), but you should not get hung up on seemingly never-ending philosophical debates. (The deduction-induction debate remains unresolved after more than two centuries of arguing!) In most fields, logical perfection is not necessary to get things done, and unless you are a philosophy student, it probably doesn't matter to your research.

Your challenge is to maintain an informed and practical—but not one-sided—attitude. You should consider deduction and induction as contrasting, yet complementary, approaches for making new discoveries and testing outcomes. Consider them—as vehicles of rationalism and empiricism—to be equally necessary intensive and extensive components in the knowledge equation. Then apply your intellectual slider accordingly.

A Diversion to the Game of Rationalism-Deduction vs. Empiricism-Induction

Here's an interesting little game that will exercise your intellectual slider and demonstrate the distinction between rationalism-deduction and empiricism-induction. It requires two people and three playing cards, one of which is a face card (i.e., jack, queen, or king) and two that are not face cards. One person, the dealer, lays the cards face down and the other person, the player, tries to guess which one is the face card. It's pretty simple, but as you'll see it's also more enlightening than you probably think; so round up some playing cards and grab a partner.

You'll play two sets of games. In the first set, one person deals and the other person plays; in the second set, the roles are reversed. Play at least twelve games per set, marking wins and losses on a score sheet like the one below. You likely will find the game most illuminating if neither of you has played it before, but whether you have played it or not, you should carefully consider what the game can reveal about the dualities in question.

	KEEP	SWITCH
WIN		
LOSE		

To begin play, the dealer shuffles the three cards and looks at the faces without allowing the player to see them; then the dealer places the cards face down in a row. Next, the player points to the card that he or she guesses to be the face card. The dealer then turns over a nonface card from the two unchosen cards. At this point, the player has another chance and may either keep his or her first choice or switch to the other face-down card. Finally, the dealer turns over the chosen card and marks the player's effort as either a win or a loss.

For our purposes, the number of "keeps" and "switches" must be equal in the match. This may be achieved either by making the first player the "keeper" and the second player the "switcher," or by allotting an equal number of keeps and switches to each player (e.g., in a full game consisting of two 12-play sets, each player would be allotted 6 keeps and 6 switches).

Before beginning play, each person should deduce the chances of the player making a correct initial choice. Will the odds change with the player's second choice, that is, after the dealer has revealed a nonface card? If so, how; if not, why not? Should the player stay with his or her first choice or switch, or doesn't it matter because once a card is revealed the odds become even for the two remaining cards? Discuss the rational-deductive logic you each use in making your decision. Once you've pondered your rationale, shift your efforts and thinking to the empirical-inductive mode and begin play.

When you've completed the full game, tally the scores and discuss the outcome from two points of view: rational-deductive and empirical-inductive. Consider how your thinking and judgment were affected, if at all, by employing both strategies. Here are some questions to consider: Did the empirical-inductive approach (i.e., playing the game) yield results that support your predetermined rational-deductive assessment and decisions, or did it cause you to reconsider them? Even if the outcome favored your rational decisions, the score undoubtedly fluctuated during the game. What do the fluctuations and final scores tell you about the precision of your experimental results as compared with your predetermined odds? If your predetermined odds were correct and definite, did they accurately foretell the frequency of making a correct choice in a given number of plays? Alternatively, if the outcome refuted your predetermined rationale, did playing the game help you identify an error in your logic?

Daydreaming vs. Systematic Inquiry

Success in research science is not the result of simply getting a bright idea, then organizing thoughts and labors along the lines of a scientific method. Success also requires imaginative inspirations, many of which are born of daydreaming. By its nature, daydreaming lacks direction. Commonly it is little more than a diversionary mental lapse—a resting stage for the mind—haphazard, frivolous, and often pleasant. But daydreaming also can be productive; it can be mental doodling that leads you somewhere.

The act of entertaining seemingly aimless musings can become an effective intellectual tool when it brings forth inspirations, releases sudden insights, and liberates ideas that might be locked in the subconscious (Ladd, 1987). The mental relaxation and uninhibited thought that come with daydreaming help us see relationships that we had not seen before. But to grasp these revelations, we first must set them up.

Consider a few imaginative thoughts and visions you've had. Were they the result of carefully reasoned thinking, or did they pop into your head? Most likely, they popped into your head, but also your head had prepared itself for the pop. No one suddenly gets an idea about possible life forms on Mars, for example, without ever having thought about Mars or about life forms. Inspirations incubate before they hatch.

One of the best ways of setting yourself up for creative insights and break-through inspirations is through systematic inquiry (which is introduced here and developed in Chapter 3). Whereas daydreaming originates spontaneously, systematic inquiry commences intentionally. It consists of carefully reasoned thinking, having purpose, direction, and discipline. Systematic inquiry can be a scheme for critically examining our fantasies and bringing ideas to fruition, but it also raises our level of awareness and stimulates the creation of those fantasies and ideas. Thus, the variety and quality of imaginative products depend on two attributes that we can systematically enhance: our knowledge and preparation.

Scientists generally are known for their ability to reason and to systematize, analyze, and verbally communicate difficult concepts—skills that seemingly originate in the brain's left hemisphere. But scientists also must be able to venture into the unknown: to learn by experience and to imagine, synthesize, and pictorially process information—activities attributed to right brain function. (This left-right hemisphere scenario is greatly oversimplified, nonetheless it serves our purpose nicely.)

You can enhance your natural talent for dreaming up new ideas, discovering creative solutions to existing problems, and exploiting your ideas by training your left and right brains to perform in concert. In so doing, your imagination becomes an intellectual slider that links the fantasy and originality of daydreaming with the realism and pragmatism of systematic inquiry (Fig. 2-6). In this regard, imagination harnesses fantasy to reality, while it frees creativity from orderly confinements. (Averill and Nunley [1992] further discuss imagination as a bridge between abstract thought and sensory experience.)

Numerous emotional, sensual, and intellectual factors stimulate our subconscious minds. While we probably have little control over most of these (and most are beyond the scope of this book), we nonetheless can set up activities and conditions to improve our productivity and effectiveness. Here are some suggestions:

- Place yourself in an environment that fosters creativity and excellence. Try to associate with interesting and challenging people. These are easy to find at most universities.

- Discuss topics—good and bad alike—freely with colleagues.

- Make times to indulge yourself in right-brain musings, and separate them from sessions of serious left-brain inquiries. Choose times and locations in which you can be free from distractions. While indulging yourself, let your consciousness stream, then free write (see Chapter 4).

- Prepare for indulgent times by writing a single word or phrase on a note card. Stare at the card to start your session.

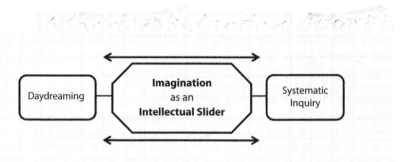

Figure 2-6. Imagination can link daydreaming and systematic inquiry.

- Be ready to capture ideas as they come to you—immediately, before you evaluate them or otherwise clutter them with complexities. For this you will need a pocket notebook or note-taking device handy at all times (even when you take a walk, sleep, and shower!). Read about free writing in Chapter 4.

- Allow diversity in your thinking, and make an effort to keep your interests and experiences varied. Insights often spring from unexpected and seemingly unrelated sources.

- Try to maintain a venturesome attitude. Accept that you can't make progress without making mistakes and that fear of error will stifle your imagination and hinder your progress. Be willing to take risks, but don't take them impulsively or with abandon.

- Allow some lightheartedness and fun into your work. Playfulness and creativity get along well together.

- Constrain uncertainty, irritation, and tension to work for you. Admittedly they cause discomfort, but in mild to moderate form they invigorate the subconscious.

- Recognize your doubts. write them down and try to resolve them systematically. When this approach seems to lead nowhere, set them aside and move on to something else. They will have been relegated to your subconscious, where your right hemisphere can work on them.

- Guard against making premature judgments and judgments based on irrelevant information or personal preferences. These can hamper your imagination. (For example, don't reject a perfectly good proposal because you don't like the person who offered it. Likewise, don't accept one without question simply because it's from someone you respect.)

- Don't always try to come up with a new concept; instead look for new links between previously unconnected concepts.

- You need dedication and perseverance, but too much continuous persistence breeds conditioned thinking. Abandon the problem temporarily to keep your mind out of a rut.

- Maintain a large account in your memory bank by rehearsing the fundamentals of your discipline and learning new advancements.

- Learn new mathematical skills and practice old ones frequently. Mathematics, including arithmetic, geometry and trigonometry, is a fundamental discipline of logic. It teaches ways of thinking to connect abstract ideas and tangible objects.

- Don't aim for a final resolution; allow room for further development of any idea.

- Don't wait for inspiration to strike—prepare yourself for it and go after it. For daydreaming to be creative, effective, and productive, you must do it and do it often. In any creative endeavor, practice stimulates more creativity.

- Strive for excellence but not perfection. According to thermodynamics, perfection is achievable only at the temperature of absolute zero, and you probably don't want to go there.

Finally, activating your imagination slider will help set you up for serendipitous discoveries—those unexpected observations, experimental mistakes, and dumb-luck happenings that contribute authenticity to your work. Chance discoveries, no matter how small or large, don't happen because luck walks in—you must lure and capture luck. So bait your senses, grease your imagination slider with a little sagacity, and prepare yourself to recognize and capitalize on any pregnant offerings that might well up. Tip the odds in your favor by coupling your knowledge and understanding with foresight and good judgment. And don't forget to practice daydreaming. To do otherwise is to try to solve problems with one brain tied behind your back.

Science vs. Advocacy

Science is not about proof in the sense of establishing validity or inducing acceptance of an idea—it's about discovery and discerning truth. This requires that the scientist be as objective and unbiased as possible. Of course, true objectivity and complete freedom from bias are not possible, simply because we are human. Our thoughts, ideas, outlooks, goals, and aspirations

Use both sides of your brain. Imagination and systematic inquiry go hand in hand. You can't create a vision, mark your path, and evaluate your progress with one brain tied behind your back.

are shaped by our upbringing, culture, experiences, and temperament. Barring cataclysmic intervention, the inborn aspects of our personalities are permanent, and childhood indoctrinations, although they can be modified, stay with us to one degree or another throughout life.

We all are culturally conditioned, and even the most disciplined among us carry part of our conditioning into everything we do, including science. Being social animals, we tend to seek out others of similar, or at least compatible, conditioning and thought style. We tend to like people who agree with us—scientists are no exception.

Despite this and other obstacles to objectivity and honesty, scientists must cultivate the ability and motivation to challenge and investigate not only the findings and assertions of others but also—and more importantly—of themselves and their kind. That scientists sometimes fail in meeting this ideal does not invalidate the ideal; it merely humanizes it.

Still, a scientist must not set out to "prove" (i.e., confirm superiority) anything; rather, he or she must set hypotheses and test them without partiality.

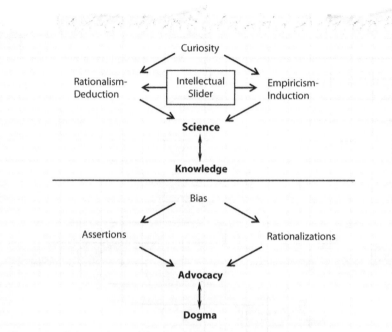

Figure 2-7. Relationships among first principles. Note the absence of an intellectual slider in the lower model.

An investigation should be able to go either way—to support or refute a hypothesis, without premeditation. To proceed otherwise encourages continued reinforcement of a single idea, which in turn produces not science, but dogma (see Fig. 2-7 for a summary of concepts).

A person who sets out to demonstrate the validity of a hypothesis is an advocate, not a scientist. While science can foster advocacy, and advocacy can generate science, the two must not cohabitate. They should be carried out by different people, independently and separately. Simply stated, premeditated advocacy presented as science is deceit.

In the discussion thus far, advocacy is considered in the context of designing and generating scientific research, not in the distinctly different context of applying research findings and endorsing scientific positions. In these latter activities, scientists are among the best-trained people to translate their work and determine its relevance to society. Nonetheless an important ethical question arises: Should or should not scientists employ their expertise to influence the outcome of public policy debates?

Although responsible scientists generally agree that premeditated advocacy is anathematic to impartial research, the question of advocacy in policy arenas stirs considerable controversy. Despite a long-standing contention that scientists should strive for neutrality lest they lose credibility, strong opposing arguments are made that scientists have a civic responsibility to offer their technical and logical skills in assessing, formulating, and promoting matters of public policy (Nelson and Vucetich, 2009).

As advocates in public debates, scientists must protect their credibility by ensuring that their positions are justified and completely open. The justification test relies heavily on dualities presented in this chapter: perhaps most important, rationalism vs. rationalization, fact vs. assumption, reductionism vs. holism, conclusion vs. inference, validity vs. soundness, correlation vs. cause-effect, and results vs. interpretation. Ultimately—amid all the weighing of alternatives, responsibilities, and beliefs—keep this advice in mind: never put yourself in a position where you can be deservedly accused of possessing the finest ethical values money can buy.

Definition Dualities: Design

Consider each of the following dualities as you design your project and write your research plan. Decide which terms best characterize what you want to accomplish. The terms in each pair are not mutually exclusive; they might overlap so both might apply, albeit to different aspects of your study or to the same aspect at different times. Also, keep these concepts in mind when reading scientific papers to see how they are used by experienced scientists. (Be aware that most scientific papers will not specify how the concepts were considered or applied; you'll have to "read between the lines" to figure it out.)

Basic vs. Applied Research

Basic, or "pure," research seeks to explain the underlying causes of phenomena. It probes for meanings without concern for immediate practical value. Scientists sometimes conduct basic research for the same reason mountain climbers take on Mount Everest—"because it's there." Although its contributions to society are not immediate, they often are huge in the long run. Basic research generates new ideas, methods, and technologies, which in turn are used by other researchers to invent new solutions to old problems. Many efforts in theoretical physics, chemistry, molecular biology, astronomy, and mathematics are basic in nature. Some examples follow:

- early work leading to the discovery of X-rays
- nature and origin of sunspots
- much of the early space research
- mapping the genome of an obscure organism

Applied research, which builds on basic research, is problem oriented. As such, projects are designed for their immediate practical importance. Most agricultural, engineering, and medical research is applied in nature. Likewise, most graduate student research is applied. Sometimes, only a faint difference separates basic from applied research, nonetheless a distinction can be important for project design, publication, and funding.

Fact vs. Assumption

A scientific *fact* is something objective and confirmed. It is accepted as true within its relevant context. Newton, for example, showed that gravitational pull can be measured by the acceleration of an object as it falls to Earth. This we consider a scientific fact. (Note that a scientific fact is not the same type of thing as an everyday fact, such as "a rock fell on my foot" or "the bus was late.")

Scientific facts commonly are thought to be discovered, and once discovered they are intransient. But are they? In the mid-1930s, Ludwik Fleck argued that despite wide acceptance of these notions, facts result from human endeavor, and as such they are created within a stratified "thought collective." In other words, when we interact intellectually and share ideas, we enter into a group whose aggregated views and direction are influenced by history, popular theories, and social conditioning. Produced in this context, scientific "facts" are not merely *discovered* independently of bias; rather, to some degree they are *created* under the influence of the collective (Fleck, 1979).

Fleck did not speak of rationalism and empiricism as distinct dualistic alternatives, but he did reveal their interaction by suggesting that rationalism, conditioned by the thought collective, guides empirical directions, interpretations, and outcomes. Although a scientific fact might begin with a "discovery," this unique event must pass scientific and societal sanctions before becoming a seemingly objective "finding."

Facts also are subject to scientific revision, as in the case of Einstein's discovery that Newton's Laws do not apply equally well throughout the Universe or near the speed of light (Hawking, 1988). (Newton's Laws still apply to our everyday lives on Earth.) So scientific facts to some degree are human constructions, which we nonetheless accept as correct. We can count on them—at least for the time being.

An *assumption* in science is something taken for granted. It is presupposed or implied and often based on anecdotal evidence; nonetheless, it is thought to have a high probability of being correct—at least by the person who makes it. Assumptions can range from previously tested background knowledge taken from a different context to arbitrary and tentative premises that are accepted without proof. Because assumptions can be comfortable, their uncertainty sometimes is overlooked. Assumptions can be important and valid in designing research, nonetheless they should be recognized and challenged before acceptance. *Assumptions cannot become facts simply by reiterating them.*

The duality facts vs. assumptions sometimes can apply to the outcome side of a project as well as to the design side.

Theoretical vs. Empirical Research

The theoretical vs. empirical duality relates closely to that of rationalism vs. empiricism. Whereas rationalism and empiricism concern our thinking modes and ideas about how we learn, theoretical vs. empirical deals more with the nature of our study. In this regard, theoretical and empirical are different approaches to designing studies or parts of studies.

Theoretical research develops, tests, or applies ideas that explain phenomena. In the first case, theory *development* studies are founded on hypotheses about the nature of a system. They can be basic or applied. They are designed to answer questions about why certain phenomena exist or happen the way they do. Most theory-development research incorporates empiricism with rationalism, but some studies rely entirely on rationalism—they are conducted without experiments or perhaps even real data. Some of Einstein's and Hawking's efforts to understand time and space fit this category. Theory-development research normally is not done at the master's level, but it might be attempted at the doctorate level.

Theory-*testing* research seeks to determine the reliability of proposed explanations or equations. Except for work in a few fields (e.g., theoretical physics and mathematics), these studies usually have a significant empirical component. Experiments are conducted or observations are made in attempts to falsify a proposed theory. For example, field-plot sampling might be employed to test a quantitative model of crop production. (Recall also the points about verification and falsification presented previously under "Deduction vs. Induction.")

Theory-*application* studies can be similar to theory-testing studies except that in theory application, emphasis is on the phenomena or desired end rather than on the theory. Research draws on underlying principles to explain observed conditions, events, or trends or to produce a desired outcome. For example, biogeochemistry can help explain the role of greenhouse gases, and aerodynamics can serve to design a new airplane wing.

Empirical studies rely heavily on observations and/or experiments to discover, characterize, and explain phenomena. Observational and descriptive studies can characterize conditions, events, and systems, but by themselves they cannot explain the causes or functioning of the observed phenomena. For this, the study must have experimental and theoretical components.

Experiments are hypothesis driven. They employ trial-and-error techniques to test hypotheses and reveal relationships among variables. The researcher must ensure that experiments are properly planned, sufficiently rigorous, and adequately replicated; otherwise, the study will not be trustworthy and doubt likely will be cast on the researcher's credibility. Although anecdotal evidence—that is, second-hand or casually collected subjective data—can stimulate good hypotheses, it is not considered scientifically sound. In all cases, empirically dominated studies must include plans for systematic interpretation and evaluation of the data.

Here are some examples of topics that might be investigated empirically:

- bird feeding habits
- shoppers' spending habits
- tree response to fertilizer applications
- psychological side effects of pain killers

Most modern research employs a combination of theoretical and empirical approaches. In your work, you should distinguish these components through all stages—planning, implementing, and closing. As with rationalism vs. empiricism, you should exercise an intellectual slider to maintain balance and ensure that your work is defensible.

Whereas a study can be wholly theoretical, it should not be wholly empirical. That is, it must go beyond mere collection and presentation of data. Data cannot speak for themselves—they must be analyzed and interpreted with respect to a theoretical component, and that's your job.

Analysis vs. Synthesis

Analysis is the major means of systematic investigation. It is a reductive and deductive process by which a system is progressively separated into increasingly specific components and studied to reveal not only its inner workings but also its overall structure and functioning. *Synthesis,* which is more constructive and inductive than deductive, is a process by which components are assembled into a new, more complex system.

Each type of activity follows a hierarchal progression, with analysis running generally down the hierarchy and synthesis running up. You can use an intellectual slider horizontally to relate and distinguish analytic and synthetic approaches, then apply another slider vertically to move up and down within either scheme. (The use of a vertical intellectual slider is explained more fully in Chapter 3.)

Analysis and synthesis can be either physical or intellectual activities, hence their components can be either material or abstract. Methods and materials of physical analysis and synthesis vary greatly among scientific disciplines. Methods include analytical procedures for chemical assays and mechanical testing and synthetic schemes for product development. For these types of activities, analytical techniques usually are more clearly defined and standardized than are synthetic procedures. Many of the principles, methods, and materials for these types of activities are explained in texts and manuals written specifically for individual disciplines. These manuals become increasingly sophisticated as the discipline develops. (Note that once synthetic procedures have been developed and are used to manufacture a product, the fabrication can be exacting but no longer is research.)

In contrast to physical practices, approaches to intellectual analysis and synthesis are more similar among disciplines—although they can vary greatly among individual researchers. They include the largely deductive thought processes involved in analyzing, testing, and applying a theory or model, and the more inductive processes used to synthesize or develop the theory or model. Additionally, the logic of intellectual analysis and synthesis can be mathematical and quantitative, or nonmathematical and qualitative.

In many intellectual endeavors, deductive analysis is more exacting and certain than inductive synthesis. Also, the path of analysis usually is more easily traced than is that of synthesis. Because this book deals with research planning rather than research methodology (as was pointed out in the preface), its focus is on intellectual, rather than on physical, activities.

Reductionism vs. Holism

Reductionism vs. holism is related to analysis vs. synthesis. *Reductionism* is similar to analysis in that the scientist disassembles a system to discern the character, function, and interaction of its individual parts, which when recombined clarify the character and function of the whole. Analysis at specific levels is an important part of reductionism, but reductionism can attempt to reveal the fundaments of the system without necessarily considering all intermediate components. To help understand a complex organism, for example, scientists can analyze it at the cellular level without studying the intermediary levels of individual organs. Accordingly, the full hierarchical progression is abridged and attention focuses only on what is thought to be the naked core of the system.

Reductionism in this sense, sometimes called empirical reductionism, serves science well. But the term carries another, somewhat belittling, meaning when it denotes "fundamentalist reductionism" (Woese, 2004). In this latter sense, reductionism connotes a metaphysical allusion of misguided oversimplification—that is, the nature of a system, including an organism, can be understood merely by piecing together component attributes without regard for synergistic interactions.

In contrast to reductionism, *holism* is a mode of thought by which a system or phenomenon is considered in its entirety. From a strictly holistic view, a system's components are so intimately related and interdependent that they lose relevance when the complete entity is reduced to simpler forms. The interaction of components creates a synergy, which gives rise to the well-known Aristotelian phrase "the whole is greater than the sum of its parts." This consideration of a system *as it exists* also distinguishes holism from synthesis, by which a system is *progressively developed* from individual parts.

Fundamentalist reductionism and holism sometimes are pitted against each other philosophically. In these cases, reductionism is viewed in an extreme, depreciative sense as a tendency to explain complex systems or phenomena in deceptively oversimplified, inadequate terms. Devotees of holism might see their viewpoint as superior and argue that attempts to break a system into its component parts and seek their separate functions is merely a frivolous or even perverse activity. In opposition, reductionists might argue that holism never would have carried us out of the Dark Ages. Holistic approaches too often are themselves overly simplistic or naïve, and that without reductionism we wouldn't have such vital fields as biochemistry, genetics, and particle physics.

Usually, such polarized disputes are no more than B-grade theater; they have little practical value because they are irresoluble (especially when actors don't distinguish empirical from fundamentalist reductionism). Reductionism and holism each have merit by offering contrasting points of view. As with the other dualities, you should exercise your intellectual slider between the two extremes, even though your studies might favor one over the other.

Finally, in designing scientific projects, you probably will find that the duality analysis vs. synthesis will yield more definite returns than will reductionism vs. holism. But keep in mind that empirical reductionism and analysis go hand in hand to produce results, while holistic thinking can help us recognize and appreciate a broader perspective.

Analytical vs. Predictive Studies

An *analytical* study breaks something—a problem, a question, a system—into its parts and then examines those parts and their interactions to discover the nature of the whole. Analytical questions usually ask How? or Why? How does something work, or perhaps why doesn't it work? Here are some examples:

- How do caves form?
- Why do caves form in limestone?
- How or why does a particular additive increase the durability of house paint?
- Why is the grass greener over the septic tank?

A word of caution about using the word "how": sometimes it indicates an analytical question (as in, How do caves form?), but other times it does not. In the house paint example, "how" and "why" are analytical and nearly synonymous. But if the question were rephrased to read, How does a particular additive *affect* the durability of house paint? it perhaps would not be analytical. This question might merely be asking whether or not the additive is effective. The answer could require nothing more than a simple comparison, which is neither analysis nor science.

Predictive studies develop forecasting models. They are designed to make projections or predictions from known or supposed circumstances to anticipated results. They also might forecast outcomes of actions, events, or circumstances. Some examples follow:

- If the price of a commodity is altered, what will be the market response?
- If shade is altered along a river, how will water temperature be affected?
- If atmospheric pressure changes by so many units, what are the chances of rain?

Predictive studies are more synthetic than analytic. That is, they synthesize or assemble (as opposed to disassemble) objects, information, or ideas into something new. Whereas synthesis normally is more inductive, analysis is more deductive.

Commonly, analytical investigations must precede and follow a predictive study. The preliminary analysis characterizes the overall system and ascertains the functions of the various components. The components, then, can be incorporated into the predictive model. A follow-up analysis normally is needed to determine the reliability of the model. These activities can be incorporated into a single research project, or they can be separated into three distinct projects. The decision of whether to incorporate or separate depends on the complexity of the system being investigated and on the wherewithal of the research team.

Theoretical vs. Operational Definitions

A theoretical definition is a universal or general statement of meaning. Although the statement should be precise, it can have multiple applications and sub-definitions. That is, a theoretical definition in science need not be tied to methodology or to particular measurement criteria. As a result, the definition can be applied differently under different circumstances—it can have various denotations. Theoretical definitions, which can be presented as narratives or as formulas, are the type found in textbooks.

An operational, or functional, definition begins with its theoretical definition, but then qualifies the theoretical expression to fit a particular purpose, such as by specifying a method of application or measurement. Operational definitions appear in scientific journals, although they might not be presented directly as such. Usually, you must read the methods section of the paper to figure it out.

This example from chemistry will help clarify the distinction. The degree of acidity or alkalinity of a solution commonly is expressed by pH, with pH < 7 being acidic, pH 7 being neutral, and pH > 7 being alkaline. Theoretically pH is the negative logarithm of hydrogen ion activity. Everyone can agree on this definition. But measuring the pH of the same solution by different methods can produce differing results, so the technique must be specified—hence the operational definition.

Here's another example. Social scientists often want to determine people's preferences among various choices. "Preference," of course, indicates a stronger desire for one option over another. Although this simple expression might serve as a theoretical definition, it is not operational nor is it adequate for designing a study. Operationally, "preference" could be defined by the proportion of test subjects who say that they would choose one option over another. But a scientist who is aware that people's actions often do not match their words could employ additional operational criteria based on observed behavior. Finally, because preference operates on a sliding scale, a threshold proportion could be defined.

Ignorance of distinctions between theoretical and operational definitions can result in serious misunderstandings, confused comparisons, and improper interpretations.

Definition Dualities: Outcome

The following pairs pertain mostly to research outcomes and their explanations.

Conclusion vs. Inference

Conclusion and *inference* refer to the nature of decisions, judgments, or interpretations made from results. They can be drawn from data and also from premises, or even assumptions. The terms overlap in meaning but should not be used interchangeably in scientific contexts. They differ in the degree of certainty.

A conclusion is clear-cut, exact, and certain. It is arrived at deductively and tends to be final. An inference is less certain, although it still can have a high probability of being correct. It is arrived at inductively and involves conjecture. It might connote some slightness in the evidence or the possibility of an alternative interpretation, or it might rest partially on an assumption (which is a form of slightness in evidence).

(An added note: Be careful not to confuse "infer" with "imply." Whereas "infer" has to do with an interpretation or the nature of an outcome, "imply" means that something is suggested without being directly stated. The meanings are distinctly different.)

Validity vs. Soundness

With regard to outcomes, *validity* and *soundness* indicate the truth of a conclusion or an inference. They depend on a decision's logical correctness and on the verity of the premises or evidence on which the decision is based. Thus, conclusions and inferences might be *valid* or *invalid*, and they might be *sound* or *unsound*, depending on the validity and soundness of the overall argument. (Recall that the overall argument is the premise or evidence followed by the conclusion or inference—refer back to syllogisms in "Deduction vs. Induction.")

A valid decision follows logically and properly from its premises or evidence, and in a syllogism it is such that any other decision is logically impossible. Being valid does not mean that a conclusion or inference is correct. It means simply that proper logic has been used to reach a judgment, thus a decision can be valid even if it is based on faulty evidence.

An invalid decision is one that violates logical standards such that even if the premises are true, the decision could be either true or false. (Note that if logical standards are violated, the decision still could be correct because of a lucky guess but not because of proper logic.)

A sound conclusion or inference requires two conditions: (i) it must be valid and (ii) it must be based on true premises or evidence. In contrast, an unsound decision is either invalid or based on faulty premises or both. By these definitions, a decision could be valid but unsound (i.e., wrong) if it logically follows from faulty evidence. If the decision follows properly from good evidence, it is valid and sound.

Here are three related examples:

1. **Premise:** The presence of plankton decreases light penetration into sea water.

   Evidence: This sea water sample contains plankton.

   Conclusion: Light will not penetrate as deeply into this water as it would if the water had no plankton.

   Assessment: The conclusion, which in this case is a deduction, is valid and sound. (Note also that the evidence statement serves as a minor premise.)

2. **Premise:** The presence of plankton decreases light penetration into sea water.

   Evidence: Light does not penetrate well into this sea water sample.
   Inference: This sea water contains plankton.

   Assessment: The inference is invalid and unsound, even though it might be correct. The limited light penetration could be caused by plankton or by something else suspended in the water. We don't know.

3. **Premise:** Light penetration into sea water can induce plankton growth.

   Evidence: Light penetrates well into this sea water sample

   Inference: Plankton will begin to grow in this sea water

Assessment: The inference is valid based on what we know from the premise and evidence; it follows logically from them. Nonetheless it is unsound because the premise and evidence are incomplete. Although some components of plankton require light, they also require other growth factors including nutrients and favorable temperature. We know nothing about the status of these additional requirements. (Note also that "light" has not been defined operationally with regard to its wavelength or other properties.)

The relationships presented in these three examples prompt numerous questions, the answers to which could either support or refute the validity and soundness of the interpretations. For example, Is the sample adequate? Is it statistically representative, and was it collected, handled, and stored properly? Is the laboratory equipment reliable and correctly calibrated for each analysis? A negative answer to any of these questions should alert us to the possibility of a valid but unsound decision.

Incongruities between validity and soundness can crop up in social science studies based on public opinion polls. These studies demand exacting protocols for design and interpretation. And of course a courtroom jury that can't recognize an incongruity between validity and soundness of evidence could send the wrong person to jail.

The duality of validity vs. soundness emphasizes the importance of scientific rigor in physical and mental activities. Sometimes decisions or judgments that seem valid and sound when they are made later prove faulty because rigorous scientific process was violated or short-circuited. The best means of ensuring that the end product of your work will be valid and sound is to use your intellectual slider diligently, plan your work thoroughly and carefully, employ appropriate scientific methods, and have your work reviewed. Also, when reviewing reports, you should read and evaluate the definitions, methods, and results—not just the "bottom line."

Correlation vs. Cause-Effect

If two variables are mutually or reciprocally related, they are said to be *correlated*. Unless the correlation is purely random (i.e., by chance alone), the strength of the correlation indicates the strength of the relationship. With quantitative data, the correlation often is depicted graphically, with the independent data plotted on the x-axis and the dependent data plotted on the y-axis (Fig. 2-8). (Note that for some engineering applications the presentation might be reversed, with independent values on the y-axis and dependent values on the x-axis.) Usually, the strength of the correlation is calculated statistically.

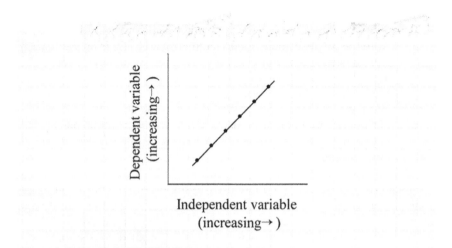

Figure 2-8. Positive correlation between two variables.

Simply because one data set, or variable, is termed independent and the other called dependent for graphical purposes does not necessarily mean that one set of conditions controls or causes the other. For example, if a person wakes up every morning suffering from a backache, we would not infer that sleeping causes backaches, even though the two conditions are correlated. Instead, we use information revealed by the correlation to guide further investigations, perhaps into mattress designs, back-muscle exercises, or even into symptoms of kidney infections.

Cause-effect relationships also are correlations, but in these cases the dependent response is shown to be caused by the independent factor. For example, the occurrence of a disease and symptoms of the disease not only are correlated, they also hold a cause-effect relationship.

Empirical studies often reveal correlations, and they might *suggest* cause-effect relationships, but they *do not establish* cause-effect. Theoretical and analytical approaches are needed to find underlying causes of observed phenomena. Figure 2-9 shows a stylized example of an inverse correlation that *might* suggest cause and effect.

Figure 2-9 is invented but it's based on real situations of soil contamination by lead from automobile exhaust, industrial emissions, and other sources. The curve reveals a correlation—that is, plant growth decreases as lead concentration increases. Is this a cause-effect relationship? In other words, is lead toxic to plants?

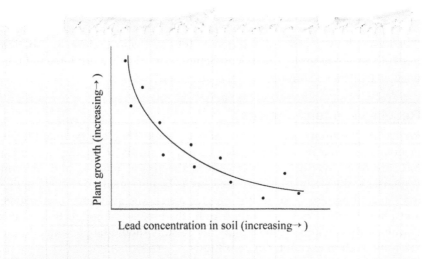

Figure 2-9. Example of an inverse correlation between lead in soil and plant growth.

The answer is possibly, but we have no way of knowing from the information presented. Lead could be the culprit, or it could be an innocent bystander, or even a not-so-innocent bystander. More studies, some using different approaches, would be needed to determine the answer. Here are some possibilities:

- Conduct additional empirical studies and analyze the soils for other potentially harmful constituents, cadmium for example. Then correlate cadmium to plant growth. Also correlate cadmium to lead concentrations. A negative correlation between cadmium and plant growth combined with a positive correlation between lead and cadmium could shift attention toward cadmium, but we still would not know the real cause of the problem.

- Conduct an analytical study of the plant's physiology to determine if lead (and cadmium) interferes with cell division or another growth process.

- Look for interactions between the metal and plant nutrients. Perhaps the metal ties up nutrients, rendering them unavailable and leading to a nutrient deficiency.

- Conduct additional correlative, analytical, and theoretical studies to determine if the heavy metals disrupt some symbiotic relationship necessary for plant health. Perhaps the metal damages a symbiont without harming the plants directly.

Note that in the above example of lead and plant growth, you would have little idea of how to proceed and how to interpret your findings without an in-depth search of up-to-date scientific literature. This is a fundamental requisite in all fields of inquiry.

Results vs. Interpretations

Results are the immediate outcome of your work. They can be quantitative or qualitative and in any of a great variety of forms, including numerical measurements, qualitative observations, theoretical models, and opinion responses to questionnaires. Results also can be presented in a variety of formats, including tables, bar or line graphs, pie charts, qualitative statements, and photographs. They are entities unto themselves: although they might be values, they hold no value judgments. Results should be presented in the "Results" section or chapter of your thesis.

Interpretations evaluate the results and explain what they mean. They point out and clarify whatever is not immediately obvious. They explain how results relate to each other and to the rest of the world, they suggest consequences, and they enlighten the reader about the significance of the work. Each result might have multiple, and sometimes opposing, interpretations. Commonly, interpretations include value judgments; for that reason, they are as much a function of the interpreter as they are of the results. Interpretations should be presented in the "Discussion" section of your thesis.

The distinction between results and interpretations appears unambiguous; nonetheless, learning to make the distinction is one of the most difficult skills to master.

For each of us, results and interpretations, or consequences, in one form or another have been intermingled since early childhood. As children, we are conditioned to immediately place a value of good or bad on numerous types of results, and sometimes we are taught to make judgments based on results we've never encountered. For example, if a toddler is warned to stay away from a stove because it is hot, the judgment precedes any result; and the toddler learns to avoid the stove even if the burner is off. (But then—fortunately for the species, but perhaps not for the individual—our curiosity prods us to experiment on our own.)

Although this early conditioning to link result and interpretation can be important for our survival into adulthood, it limits our creativity, and it inhibits our ability to make useful distinctions later in life. Scientists, like the rest of the population, are faced with biases every day. They continually must guard against unduly coloring their work. While some biases relate to the science, as in adherence to paradigms and preconceived notions of outcome, others stem from social, cultural, moral, political, and religious beliefs. A few even arise from ill-advised peer pressure and financial concerns.

Certainly, a scientist can have beliefs apart from science, but he or she must exercise diligence beyond the norm in designing studies, collecting data, and interpreting results. *An important part of any graduate program should be to learn to recognize personal or societal conditioning and to go beyond it.* Without this type of personal breakthrough, distinctions remain blurred, options remain limited, and vision reaches only to the immediate horizon.

(A cautionary footnote: Beware of pronouncements of pseudoscientists and other advocates who knowingly and skillfully fuse results and interpretations to further some purpose. These deceptions are especially difficult to recognize when the purpose is one in which we believe.)

A Triad: The Testable, the Trusted, and the True

The terms *hypothesis, theory*, and *law* epitomize much of research science. In the scientific context, the words are used differently from their use in popular writing and everyday conversation. Scientifically, they usually denote a ranking in certainty of concepts, but even within the scientific community, they can have somewhat different meanings to different people. The distinction between theory and law can be particularly fuzzy and confusing—so much so that we might even question the utility of using both terms when one might suffice. Nevertheless, both are in common use, so they will be addressed here. Rather than debate their meanings, we simply will define and employ them in the following ways.

Hypothesis

Hypotheses are carefully conceived speculations that guide research. Although they are tentative and uncertain, they are more than wild guesses or rationalized conjecture. They deal with something that is perceived or suspected on the basis of limited evidence and rational construction but not yet known. As conceptual tools, they can take many forms and have many applications, including ideas or proposals about the nature of an entity, a system, an event, or another idea. Hypotheses offer paths for inquiry when planning observations and designing experiments.

Bear in mind that a hypothesis is a tool—a means to an end, not the end itself. With hypotheses, scientists should be tool fabricators, testers, and users; not tool collectors, custodians, or sellers. They should avoid becoming hypothesis merchants who try to convince others of the excellence of their wares. If a tool proves inadequate for completing a job, the scientist should set it aside and take up another.

What a hypothesis is not: a scientific hypothesis is not a simple guess or assertion about some verifiable set of characteristics, conditions, or events. For example, guessing the number of beans in a jar, predicting the outcome of a sporting event, and forecasting the weather do not engender hypotheses. Verification in the first example requires nothing more than a trivial counting exercise. The second example entails study of past performances and current conditions, but in the end we need only note the outcome of a single event. The third example employs established scientific principles, but applying the principles to current conditions does not require a new hypothesis. Each of these examples involves a guess or projection about something that is finite, specified, and amenable to full measurement within a delimited time span.

A scientific hypothesis must direct the investigator beyond the knowns of a narrowly defined sample to the unknowns of a broader population. A valid hypothesis, then, must do more than provide guidance for accumulating facts or enumerating data: it should lead to expanded knowledge. So while a weather forecast is not a hypothesis, a reasoned speculation about how atmospheric parameters interact to create weather could be. (Note: Nothing precludes a hypothesis from being specific or narrowly defined—in most cases it should be—but also it should make expanded knowledge possible.)

Hypotheses must be tested, and testing must be planned. This is an important part of experimental design and research methods. As emphasized earlier (see previous sections "Deduction vs. Induction" and "Science vs. Advocacy"), *a scientist—you—never should set out to confirm a hypothesis.* Rather you should test it honestly and rigorously. If the hypothesis proves wrong, so be it. As the 19th-century biologist Thomas Henry Huxley ([1900] 2006) noted, "Next to being right in this world, the best of all things is to be clearly and definitely wrong, because you will come out somewhere" (p. 249).[1] Also, you will be far better off if you prove your hypothesis wrong before someone else does.

Upon testing, most hypotheses fail in some measure. They begin life as seemingly bright ideas and end as lowly—but nonetheless, useful—negative results. Scientists, despite their training to falsify hypotheses, naturally tend to prefer resilient ideas, and when an idea they've spawned and nurtured proves faulty, the impact can be harsh. As Huxley ([1894] 2004) remarked, "the great tragedy of Science—a beautiful hypothesis slain by an ugly fact" (p. 244).[2]

---

1 Along these same lines, but in an uncertain context, the 20th-century physicist Wolfgang Pauli famously remarked something like this: "This theory is useless. Not only is it not right, it's not even wrong!"

2 Although this aphorism (with slight variations) appears in Huxley's writings, it also is attributed to Friedrich Wöhler in a communication to Jöns Jakob Berzelius shortly after Wöhler had synthesized urea in 1828, when Huxley was but a toddler.

Putting Huxley's two quotations together, one might perceive that the author was contradicting himself. Moreover, when we recall that (i) a hypothesis cannot be inductively proved and (ii) a hypothesis is not scientific if in principle it cannot be deductively disproved, we might consider that one of the *fortunes* of science is a slain hypothesis. The "tragedy" then is of personal ego, not of science.

Falsification is emphasized over verification when stating and testing hypotheses for at least two reasons: first, this technique complies with a logical principle, and second, it helps regulate the common human tendency to seek and accept solutions that quickly satisfy the curiosity (Medawar, 1967). We humans are prone to accepting plausible explanations: if an explanation fits the situation and seems reasonable, it's often good enough. In science, we must go beyond that tendency.

Through rigorous testing, we usually discover that a phenomenon has more than one conceivable explanation. Accordingly, a single problem can support multiple hypotheses. The facts and data we collect are relevant or irrelevant with respect to the hypothesis, not to the problem itself.

Studies involving the collection of numerical data normally are statistically designed, and the data are subjected to statistical testing. The process involves contradictory conjectures called an initial, or null, hypothesis and an alternate hypothesis. The tests usually are conducted not to choose the better of the two statements but to determine which one is incorrect. Statistical procedures are not guaranteed error free, but they can give a probability of making an error.

Some hypotheses withstand the tests. They emerge adequately reliable, guide researchers, and lead to new discoveries. Occasionally, one will pass multiple comprehensive trials and rise above all competitors as scientists promote it to theory status (at least in a de facto if not a de jure sense).

Theory

Whereas hypotheses set the direction for projects conducted by individuals or small groups, theories can set the direction for entire disciplines or even societies. A theory is a trusted and reliable statement of knowledge—a basis for understanding—that withstands intense scrutiny.[3] Compelling logic and evidence have shown it to be valid and sound under a wide variety of circumstances. It develops rationally and empirically from either a single hypothesis or an aggregate of related hypotheses. Although the criteria are demanding, they do not require that all aspects of the theory be resolved and understood. A

---

3 Be careful not to confuse "theory" with "theorem," which is a mathematical construct built on or provable by a set of axioms.

theory remains somewhat provisional. Adjustments are allowed; nonetheless, the basic premise serves as a useful predictive model and is widely accepted.

A theory should not be confused with a fact. Whereas a fact might refer to what happens, a theory explains how it happens. Evolution, for example, happens. It is a fact. Explanations about *how* and *why* it happens are theories (Gould, 1983). Another example: We can take the existence of gravity as fact. Why gravity happens is anybody's guess.

Theories, which can be mathematical or nonmathematical, are proposed from either rational deduction or empirical induction. If a proposal proves deficient in either logical form, it fails the test as a legitimate theory, at least in the scientific sense. Here are a few examples, each of which would begin as a hypothesis:

- Overwhelming evidence shows that people who eat broccoli eventually die. Based on this evidence, one might propose the theory that eating broccoli shortens the human life span. Obviously, the premise—that people who eat broccoli eventually die—is true, but the inference is invalid and unsound. The interpretation depends solely on a correlation, with no cause-effect evidence, and considers no other possible causes of death. The proposed theory fails the rationality test. (Note that this example suggests nothing beyond a normal diet that includes broccoli. Eating rotten, contaminated, or grossly excessive amounts of broccoli might cause health problems, but that's a different premise.)

- Two millennia ago, Ptolemy, the Alexandrian astronomer, theorized that the Sun and other heavenly bodies revolved in concentric circles around Earth. Presumably, his theory was deduced from philosophical (perhaps religious) beliefs that humans live at the center of the universe. It was supported by empirical evidence—he could see the sun rise in the east, pass across the sky, and set in the west.

    Ptolemy's theory ran counter to a proposal made three centuries earlier by Aristarchus, a Greek, who thought that Earth and the other planets revolved around the sun. Although Aristarchus's interpretation was correct and Ptolemy's was wrong, Ptolemy's notion fit accepted rationalizations of his time—it was "common sense." The Ptolemaic concept reigned for 15 centuries, until Nicolaus Copernicus convinced astronomers that the sun is the center of our solar system. Subsequently, Copernicus's theory was proven (with modifications for elliptical rather than circular orbits) by Johannes Kepler, who employed mathematical rationalisms to confirm empirical evidence (Bunch and Hellermans, 2004).

- Wildfires, which are common events in many ecosystems, can lead to accelerated soil erosion on steep hill sides. For decades, land managers have artificially seeded burned slopes with introduced grasses to quickly establish plant cover and thereby control erosion faster than would happen through natural recovery. The hypothesis (i.e., rationalization because it was not adequately tested) that artificial seeding would control erosion was quickly accepted when blackened hillsides turned green with the seeded grass (i.e., empirical evidence).

  Although the argument favoring blanket seeding seemed valid, its soundness eventually proved questionable when actual erosion measurements revealed that introduced grass does not necessarily control erosion. Instead, it can interfere with natural plant recovery and sometimes stimulate unwanted animal activity—both of which can lead to more, rather than less, erosion (Taskey et al., 1989).

  In this example, a rational but unsound premise persisted in accepted practice for several decades. During that time, observations repeatedly showed that artificial seeding caused rapid increases in plant cover, but concurrent direct and statistically valid measurements of actual erosion were not made. Instead, most managers accepted the common assumption that artificially increased plant cover always controls erosion.

  In effect, a hypothesis had been accepted and elevated to theory status without proper empirical evidence because it "made sense." Subsequent management decisions were justified merely by rationalizations. (Note: A favorable cause-effect relationship between plant cover and soil erosion holds in most situations, but it has limitations.)

- Here's an example of a different sort—a political rationalization posed and perpetuated as theory but one lacking any scientific basis. During the U.S.–Vietnam War, the "domino theory" was given as a major reason for U.S. involvement. Proponents of the "theory" maintained that if the United States pulled out of Vietnam, other countries in Southeast Asia would fall one by one to Communism and that would be bad. Although the argument was plausible and accepted by millions of Americans, it was never supported by empirical evidence. (In fact, less than two decades after U.S. withdrawal, Communist governments collapsed in much of Asia and Europe.) The proposed theory failed the empirical test, but only after tens of thousands of people had lost their lives.

As in each of these examples, theories normally encompass occurrences and mechanisms. Sometimes occurrences are palpable, but other times they are only hypothesized. A genuine but unseen hypothetical occurrence will be strongly doubted if it lacks a convincing mechanism. Conversely, an occurrence worthy of severe doubt might be widely accepted if it is backed by a believable mechanism. Once researchers confirm an occurrence, they continue to challenge its proposed cause which may be driven by still other causes.

Plate tectonics offers an excellent example. In the early 19th century, geologists noted similarities in fossils and rock types found along North America's eastern edge and Europe's western edge. Over the years, geologists discovered more overlapping phenomena and searched for explanations. About 1910, the German geologist Alfred Wegener suggested that the two continents once had been joined, later split apart, and eventually drifted to their current locations under Earth's centrifugal force and solar gravitational pull (Hurley, 1968). "Absurd," muttered skeptics, "common sense says there's no way entire continents can float through solid earth!"

Another half-century passed before studies revealed that Earth's crust consists of an assemblage of huge movable plates, and that lava spews from a great plate-spreading rift in the mid-Atlantic floor. Seafloor spreading might drive continental drift, but what drives seafloor spreading? The answer lies in complex mechanisms of convective thermal upwellings from Earth's mantle (Plummer et al., 2007). And how do these upwellings function? For answers, scientists continue to pursue occurrence-mechanism, or effect-cause, relationships.

Law

Eventually if a theory is sufficiently verified and far-reaching, it can be accepted as a scientific law. It must have withstood intense rational assessment and rigorous empirical testing. Laws commonly can be expressed in equation form, which allows them to be applied mathematically.

Despite their universal acceptance, laws do not necessarily apply universally. They apply within a defined realm. In other words, laws operate effectively within a range of reliability. Their precision diminishes toward the fringes of that range; nonetheless, they yield extremely useful and reliable results within it.

The defined realm might not be fully recognized at the time of a law's acceptance. For example, Newton's Laws once were thought to be universally valid. Although they produce highly reliable results when applied to macroscopic objects within Earth's gravitational field, they fall short at atomic and stellar scales and at light-speed velocities. Still, apples continue to fall from trees just as Newton observed, and not even his laws can be expected to surpass their design limitations. Likewise, Boyle's Law works well at the low pressures under which it was formulated, but scientists later discovered that its reliability diminishes with increasing pressures, which Boyle's work never could have attained.

Given the above definitions of theory and law, one might assume that theories can only precede laws, but theories also can follow laws if they are developed to explain the law or expand its realm. For example, Newton's Law of Gravity describes only *how* bodies behave in Earth's gravitational field. It suggests nothing about *why* they behave the way they do. Additional theories are needed for that.

The foregoing discussion suggests that a theory or law will be accepted if it withstands rigorous scientific testing. But there's a catch: it also must escape significant societal controversy. A theory that scientifically might qualify as a law could be denied that status if it runs counter to strongly held societal beliefs. Science does not stand alone in the world.

And Then There's the Paradigm

By its simplest definition, a paradigm is a model or example that sets a pattern. In science, it's akin to a theory, but it's also more far-reaching. Like a theory, a paradigm is a tested and trusted framework that helps set intellectual direction for a discipline. But compared to a "normal" theory, a paradigmatic theory is stronger and overall more compelling to believers. In going beyond, a paradigm also encompasses the theory's attendant facts and modes of operation. It even takes in the collective mindset of scientists who work within it (Kuhn, 1996).

A new paradigmatic theory (or group of related theories) is sufficiently intriguing and superior to the earlier controlling theory that it attracts fresh adherents and begins to supplant earlier direction. The replacement theory also is sufficiently different and open-ended that adherents have ample opportunity to contribute and build entire careers working within it (Kuhn, 1996).

A paradigm's significance is great enough that the process of change—called a paradigm shift—constitutes a scientific revolution, at least in the affected discipline (Kuhn, 1996). The shift engenders not only new ideas and opportunities but also new methods, new technologies, new sources of funding, and new textbooks. In some fields, a shift can be of such consequence that its effects ripple throughout society. Note for example the impact of Newton's work, Einstein's contributions, and the dawn of the space age. Paradigm shifts move science forward.

Improvements and benefits notwithstanding, paradigms are difficult to change. For most scientists, their careers are born and raised in the community of an existing paradigm. Although they've had special training in innovative thought, scientists remain people—to varying degrees, they are subject to conditioning, bound to tradition, and trustful of the ways of their mentors. Most are not rebellious, and to them an impending revolution can be upsetting, even traumatic. It threatens their work as well as trusted practices. To them, the proposed change might portend a degradation of their achievements and a devaluation of their contributions. No one wants to see his or her life's work become passé and relegated to meaningless history.

In addition to disrupting the status quo, paradigm shifts can be costly. The cost list is lengthy, but the following examples will give you a sense: reeducating established scientists, retooling laboratories and updating educational facilities, retraining technicians and hiring new ones, updating educational curricula, rewriting and publishing text books, expanding scientific journals, reevaluating and revising past work, and creating new standards and rules for the field, in some cases for ethical and regulatory purposes.

Given all the obstacles, what causes scientists to initiate paradigmatic change? Can't they find enough work and reward in "normal science"—that demanding and tremendously productive pursuit that can run for decades following a major shift? Most scientists can. They spend full, satisfying careers making discoveries and advancing their profession. But accidents happen: an experiment runs amok and produces an unimagined result. Crises erupt: political turmoil or an epidemic demands immediate attention and a revolutionary approach.

Other reasons are more common, albeit less obvious and dramatic. Crises of a different sort can develop insidiously: long-term, continued testing can reveal unexplainable gaps and weaknesses in theory and practice. Eventually these become insurmountable and impossible to dismiss. They lead to "technical breakdowns" and gradual loss of confidence in the paradigm. A few ambitious researchers will feel forced to abandon current direction and find a totally new approach (Kuhn, 1996).

In other cases, anomalies will appear more repeatedly and persistently as technology advances, the research field expands, and the paradigm matures. Persistent anomalies can breed frustration because they suggest failed efforts, but they also can signal opportunity and the need for change. In time they must be dealt with head-on, and new mechanisms must be found to explain their occurrence. The result can be a paradigm shift (Kuhn, 1996; Woese, 2004).

Paradigms are extremely resilient. They should be—they provide a foundation for scientific advancement. Science cannot launch from thin air—it must have a base. So when one paradigm shows signs of wear, another must be proposed to take its place. Then the two must be matched one against the other for supremacy. In the end, the scientist can feel confidence in the new paradigm. Confidence is vital. Without it, a scientist cannot focus, cannot synthesize, and cannot ferret out the details that make things work.

The case of continental drift and plate tectonics as occurrence-mechanism theory development (see "Theory," earlier in this chapter) superbly exemplifies a major 20th-century paradigm advance. Check out Stephen Jay Gould's timeless little essay "The Validation of Continental Drift" (1977). Gould deftly reframes the case for the "new orthodoxy" without ever mentioning the word "paradigm." He taps in critical bits of epistemology, logic, and scientific philosophy, but never hits you over the head with them. No matter what your field, it's worth reading.

The "tree of life" provides another far-reaching example. In the old days, life forms were classified into two kingdoms: plants and animals. Biologists recognized early on that this system was inadequate, and gradually they expanded it to five kingdoms. But still, the system remained phylogenetically deficient, partly because it discounted the most widespread, prolific, and diverse organisms on the planet: the microbes.

In late 20th century, a small group of innovative biologists addressed this deficiency by employing cladistics to radically restructure the system of biological classification (Woese et al., 1990). Their proposed phylogenetic tree (or cladogram) provides a combined taxonomic and evolutionary framework for classifying organisms. It is based not on what organisms look like but rather on their molecular make-up (e.g., ribosomal RNA), which can reveal ancestral linkages. The tree supports three main branches, or domains: bacteria, archaea, and eucarya (also spelled "eukarya"). On this tree, animals (including humans) and green plants no longer branch equally from the stem of life; rather, they sprout, along with a few other groups, as offshoots from the eucaryotic branch.

So why should a rearrangement of a classification system rate the distinction of a major paradigm shift in science? Answer: Because it's more than a taxonomic rearrangement. It's a replacement hierarchy—one that reorganized knowledge, redirected thinking, and redoubled efforts toward genetic and evolutionary understanding. It, along with the genetic sequencing that led to it, prompted a reassessment of biological diversity and evolutionary understanding (Woese, 2004; Pace, 2006).

A new paradigm brings new theories, new protocols, and a new thought collective. It brings excitement and opportunity and invigorates the pursuit of knowledge and truth. People—even devout nonscientists—are naturally drawn to new paradigms. Some newcomers bring hot new blood. They are eager to attack problems and ultimately make a name for themselves. Others contribute greatly but more discreetly. Part of their reward is being able to say "I was there when . . . ."

As a graduate student, don't bother trying to effect a paradigm shift. A new paradigm can take years to become accepted—sometimes only after the proponents of the older paradigm pass on. Operating paradigms are bound by the tradition of their proponents, who will defend them as if they were the castle walls. So stick to "normal science." Pick a puzzle to solve; puzzles have solutions. This advice is especially relevant for master's students. Alternatively, if you are an ambitious, adventurous, and hardheaded doctoral student with a competent and supportive graduate committee, you might want to go for it—a few graduate students have been remarkable shakers and shifters of paradigms.

Resiliency of paradigms perhaps is part of a natural system of checks and balances. The system helps us avoid headlong, costly, and sometimes disastrous ventures into the unknown. Certainly, scientists live to probe the unknown, but they do it with some degree of knowing. That's what the next chapter is about.

CHAPTER 3

Research Planning

What's This Chapter About?

Systematic inquiry: Organizing and exploiting your creative thinking

Preparing research plans and grant proposals

What's in This Chapter?

Science is, I believe, nothing but trained and organized common sense...

T.H. Huxley, On the Educational Value of the Natural History Sciences

Not quite—and undoubtedly Huxley, who was a thoughtful and competent 19th-century biologist, not a careless quipster, really knew that. Although "trained and organized common sense" are central to good science, science is driven by research; and research is impelled by curiosity, necessity, dissatisfaction, and of course profit motives. In seeking to discover and explain the unknown, successful researchers must be creative, intuitive, and inspired. But this isn't enough. To bring it all together, they also must harness their talents and direct them to a fruitful end—they must organize their genius as well as their common sense.[1]

While the dualities in Chapter 2 offer alternatives for inspiring, sorting, and directing thoughts and ideas, this chapter shows you how to exploit those thoughts and ideas through planning your overall research. Research planning demands *systematic inquiry*—a structured method for recognizing and manipulating intellectual resources, including established principles and recent findings, as well as your own insights. Although scientific research does not progress according to a single "correct" method—indeed, styles vary greatly among scientists—beginning students need a model to get started and to make timely progress. That's what this chapter is about.

The planning system we will use incorporates exploitation factors into a hierarchal framework. As you will learn, a hierarchy can be more than simply a diagram for displaying administrative rankings. When constructed diligently and used sagaciously, it becomes a powerful tool for analyzing or synthesizing a system, which is what much of scientific research is all about. Hierarchies can help isolate problems and foster ingenuity. They can be built and operated either rationally and deductively or empirically and inductively, or by a combination of approaches.

To make them work requires an operational mechanism, and once again an intellectual slider will do the job. In Chapter 2 you learned to operate intellectual sliders horizontally between contrasting choices, but a slider also can

---

1 Note also that many aspects of common sense are culturally conditioned. They depend on the values, customs, and traditions of a society, in addition to the beliefs of an individual; therefore, they can differ significantly among peoples, and they can change over time. Common sense, then, is more than an informal synonym for horse sense, which is determined more by biological instincts. (Applied to humans, horse sense might be considered a subset of common sense.) Although Huxley went on to clarify his comment, he did not address the impact of this important distinction on science, nor did he assess the role of individual temperament in applying common sense to science (Huxley, 1902).

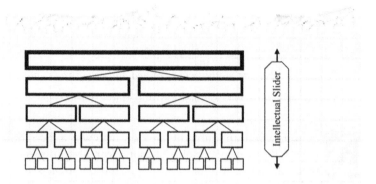

Figure 3-1. Use a vertical intellectual slider to build and operate a hierarchy, as well as to shuttle among its levels and across its scales.

function vertically among multiple choices. Applied vertically, your slider need not be limited to options of comparable consequence; rather, it can be applied across scales, as in a hierarchy (Fig. 3-1). Operated in this mode, the slider transforms the hierarchy from a static display to a virtually dynamic research tool. (A good example was presented near the end of Chapter 2, where we noted that use of cladistics, the hierarchical method of connecting taxonomy and evolution, led to major advances in biological knowledge.)

The guide (i.e., model) we will use for research planning and problem solving represents one version of the so-called scientific method (see Fig. 3-2). Although our model is hierarchical and linear, most of us do not think this way. Rather, we might entertain ourselves with right-brain daydreaming one moment and challenge ourselves through left-brain cogitating the next. Even when cogitating, our thoughts can jump about from one point to another following some sort of nonlinear logic. For this reason, you probably will start your planning somewhere in the middle of the topics in this guide. That's OK. Don't fight it—dive in and call it "brain storming." Just remember that you probably are starting in the middle, even if you think you are starting at the beginning.

Even though our thought development process might be messy, in scientific work we ultimately must present the information in an orderly sequence to ensure easy, accurate, and unambiguous comprehension by ourselves and our readers. So although you'll skip around in preparing your research plan, make sure that your final version is logically arranged. You might not need to

The Introduction

1. Title page
2. Abstract (prepared last)
3. Table of contents
4. Student's professional and academic goals
5. Introduction
   A. Background information and problem statement
   B. Statement of the overall goal
   C. Identification of subgoals
   D. Statement of subgoal to be investigated
   E. Importance of the project (Why? How? To whom?)
   F. General approach

The Middle Stages

6. Literature review
7. Key factors and variables
8. Scope
9. Objective(s), or hypotheses to be tested
10. Assumptions
11. Methods and materials
    Field and laboratory
    Questionnaires
    Statistical design
    Methods of analyzing data or interpreting results

The Closing Stages

12. Anticipated results, including tables, graphs and charts
13. References (follow format of ASA or other approved professional society)
14. Publication plans
15. Timetable
16. Budget
17. Appendices (will vary depending on the nature of the project)
    Definitions of terms
    Data sheets
    Supplies and equipment list

Figure 3-2. Research plan outline.

include each item separately in your plan, nonetheless you should think about each one. If you omit an item, you should be able to defend your decision.

Also, some points in the research plan are for planning purposes only and will be edited out of your thesis. They are analogous to scaffolds used in building construction: necessary for erecting the structure, but unnecessary and distracting to the finished product. Once you have an idea under control, you may remove its scaffolds.

From time to time, you likely will get off track and confused, especially when starting out. Don't worry about it—press on. That's all part of training and organizing your "common sense." Keep in mind that the model (i.e., the research plan outline) has a built-in system of checks and balances. This system is not infallible; nonetheless, it can help you maintain consistency and recognize when you go astray.

Although the plan outline is fairly detailed, you don't have to get carried away with details. Use the system to organize thoughts and information, but don't use it so rigidly that it stifles adventure and creativity. Still, for your first project you probably should follow the outline fairly closely. With experience, you might want to modify the model to better suit your personal style, fit a particular project, or to comply with different standards in your field. Feel free to do that, but make changes knowingly and purposefully.

Your research plan will be a dynamic document, fraught with uncertainties and subject to many changes, but packed with information, direction, and good ideas. And once you have written it, you effectively will have completed the first draft of your thesis. Step by step, you will transform it into the final copy of your thesis or publishable research paper. Do your best to shape and compose it, but feel free to express your doubts as you write. You may even include a note requesting help from your reviewers at the beginning or end of any section.

Keep your readers in mind as you write your plan. In most cases your writing should be directed at people in your discipline, some of whom will be expert and others just beginning. A good guideline is to write at a technical level that readily would be understood by other graduate students in your field but who perhaps have different specialties. Use technical terms when they are necessary, but avoid jargon—that is, words that are unnecessarily technical and pompous. Jargon comes across as pretentiousness and does little more than irritate and limit your readership.

When you ask, Where should I start? the next question should be Where should I stop? or How will I know when I'm finished? Preparing your research plan will help you define your starting point, establish your end point, and set the sideboards. It can be a challenging but stimulating and rewarding process. Learning to think and to apply your thinking in this way is one of the main reasons for being a graduate student.

Finally, keep this in mind: Even if you complete all course work with perfect grades, no diplomas are awarded for not finishing the required dissertation. Writing a research plan and updating it regularly is one of the best activities for moving you toward successful completion.

The Research Plan Explained

The Introduction

Title Page

The following title page format is fairly standard and accepted by most universities (Fig. 3-3):

Abstract

An abstract is a short, objective account of a larger work. Two types normally are recognized: *indicative* abstracts and *informative* abstracts. An indicative abstract summarizes the type of information in a paper; it tells what the paper is about or what has been done without necessarily divulging the findings. It commonly is only three to four lines long and of limited value. In contrast, an informative abstract focuses on the essence of a work by summarizing its ideas, facts, and findings, usually into a single paragraph. You should write an informative abstract for your research plan and thesis.

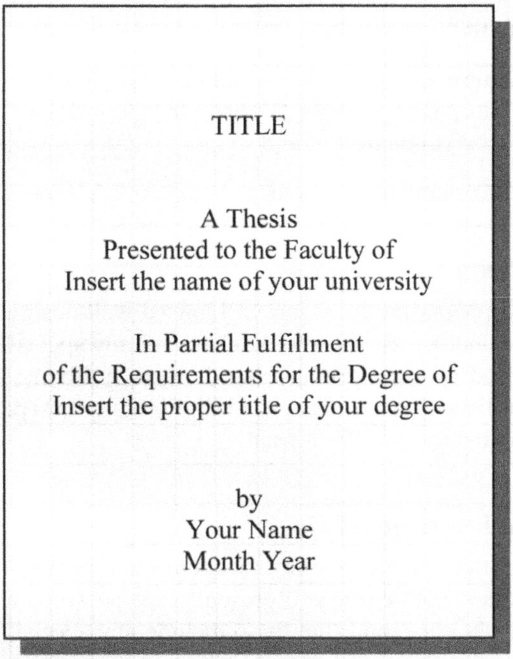

TITLE

A Thesis
Presented to the Faculty of
Insert the name of your university

In Partial Fulfillment
of the Requirements for the Degree of
Insert the proper title of your degree

by
Your Name
Month Year

Figure 3-3. The elements of the title page.

An abstract typically runs about 100 to 200 words, with a maximum length of about 300 words. For your research plan, it should contain the following information: the problem to be investigated (or goal of the project), the objective, the methods to be used, the anticipated results, and the significance of the work. It may be written in future tense or present tense, or some combination of the two when logical and appropriate. Although the abstract appears at the beginning of the research plan, thesis, or published journal article, it is written last.

The thesis abstract is similar to the research-plan abstract except that it gives actual results and includes conclusions or recommendations, and it is written in past tense. Although the abstract is part of the thesis, it can be published separately, such as in *Dissertation Abstracts* or on any of numerous internet sites. The possibility of separate, broad publication should cause you to take extra care in preparing the abstract.

The following guidelines will help you write a concise informative abstract:

- Do not repeat information given in the title.

- Write concisely (almost tersely if need be) but not in primer style.

- Write in the active voice, giving careful consideration to the choice of subject and verb. Use active verbs; avoid using the verb "to be."

- Avoid empty expressions, redundancy, and indefinite pronouns (e.g., "This paper presents," "The author found that," "due to the fact that," "There are," "It is").

- Keep the organization linear.

Table of Contents

Any good word-processing program will have a well-designed table of contents function; nonetheless, you should check for any special formatting that might be required by your graduate studies office, library, or professional journal. Present lists of tables, figures, and photographs separately from the table of contents but in a similar format.

Academic and Professional Goals

Give one brief paragraph each for your academic and professional goals. Academic goals include the degree you presently are pursuing and any degrees you're thinking about beyond that. Also include any special certifications or other achievements requiring extra education and training. For professional goals, think short-term as well as long-term. What type of work do you think you'd enjoy doing? For whom and where would you like to work—a private company, a government agency, an institute, yourself? Is international work appealing to you? If you are uncertain, that's OK; say so and identify some of the things you know you would not like to do.

These statements will be useful to your graduate committee and others who review your plan. Also, writing the goals will help you decide personal directions as well as prepare for job interviews and related activities. This also is a good time to consider how your professional goals fit with your personal goals, including family life. Do not include these academic and professional goal statements in your thesis.

Background Information and Problem Statement

This section sets the stage for your work. It provides background information readers need to grasp your research problem and goals. If your problem is commonly known, it will require little preliminary explanation. Conversely, if it is obscure or complex, you must describe it so that readers can visualize the problem and appreciate your intentions. Don't get into details; those will come later. In most cases, you can outline adequate background in a few paragraphs. Include citations from professional journals or other appropriate sources to verify the problem.

Here are some examples to start you thinking:

- Perhaps evidence suggests that muscle pain increases as people age. State that, then highlight a bit of evidence. Are some muscle groups more susceptible than others? When does the pain become debilitating, if ever? Are current treatments therapy based or drug based?

- In recent years, food production in a remote village has begun to decline. Give the village's geographic location and climatic zone. Outline some evidence and possible reasons for the decline. What are the traditional crops? Briefly describe the farming system and some of its likely limitations.

- If you're interested in preventing overheating in truck brakes, you might cite conditions under which problems occur and the types of brakes involved. You also could mention factors of heat generation and how these relate to the problem.

- Perhaps you're interested in more basic science and would like to discover the enzymes, genes, and energy transformations involved in a biochemical reaction. You might begin with an overview of the reaction and its significance, then point out gaps in the current knowledge.

Note that in giving background information, you have begun to formulate and imply hypotheses. These will be interwoven throughout the process. You will think about them before clarifying your overall goal or establishing key factors and variables. Although the section on objectives and hypotheses is placed midway through the research-plan outline presented in this chapter, you must learn to recognize and state your hypotheses as best you can whenever they become apparent to you.

Overall Goal

While the background information above sets the stage and broadly defines the problem, the overall goal expresses an approach for solving the problem. It lies at the top of the hierarchy (Fig. 3-4). It is broad based—umbrella-like—and much too large for one person to accomplish in a single project. As such, it overarches the subgoals, which are more specific, achievable ambitions. By

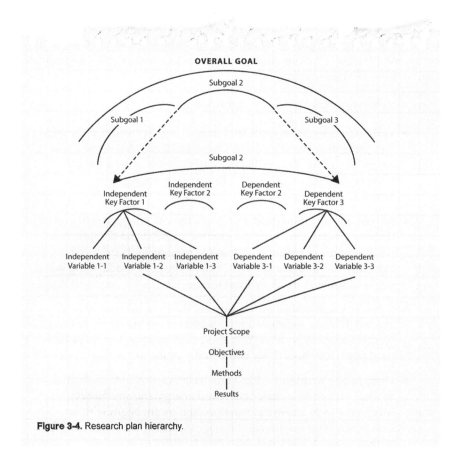

Figure 3-4. Research plan hierarchy.

setting the larger context, the overall goal helps you and your readers put your specific objectives in proper perspective.

A study's overall goal might be to alleviate muscle pain in elderly people or to develop a sustainable agriculture program for a village. These goals have many more facets than one person can investigate with limited time and resources. That doesn't matter—they set the context for more specific sub-goals. State your overall goal in a short paragraph at most.

Subgoals

Break your overall goal into several subgoals and state each as a one-liner or brief sentence. Do not elaborate at this time; that will come later. For example, an overall goal of increasing profits in a particular industry might be broken into the following subgoals: (i) increase efficiency in some aspect of production, (ii) enhance product utility, (iii) improve marketing strategy, (iv) alter financing methods, and so on. Try to arrange your subgoals into logical categories. In this example, some subgoals might be aimed at decreasing production costs, whereas others are intended to increase gross income.

Bear in mind that you are constructing a hierarchy of ideas, and in so doing you can add more tiers as necessary. You might find that a tier of sub-subgoals will be helpful—just don't overdo it. Often a better approach than adding more tiers is to reconsider your overall goal and level of background information. Note that in the above example of industry profits, any of the subgoals could be elevated to the level of overall goal. Of course, the background information and problem statement would need to be adjusted accordingly.

Note that by using this technique, you could design your project for any size group, ranging from a single graduate student to an institute or corporation. Large, multifaceted projects, such as those conducted by research groups, require multiple hierarchies. Each hierarchy is designed for a specific branch of the organization, and each displays different groupings of specificity.

In most cases, your study will be aimed at only one of the subgoals listed—that's why they're only one-liners at this time. The remaining subgoals are part of the scaffolding of your research plan. They will be edited out of your thesis or professional paper, but in the meantime they help set direction and define context. They show your reviewers that you are not overlooking important aspects of the problem and that you are putting your proposed work in proper perspective. In addition, the extra subgoals can provide alternatives should your project meet severe difficulties.

Statement of Subgoal to Be Investigated

Arrange your list of subgoals in a priority order, and select the one having highest priority for you (see the section "Next You Need a Research Topic and a Strategy," in Chapter 1). This is the part of the overall goal you will work on. Write it in a clear, concise statement or brief paragraph. This statement or a revised version will be included in your thesis.

Note that the highest priority to you might not be the highest priority to the rest of society. That often doesn't matter either. For example, if the greatest challenge to increasing apple-farming profits is to improve marketing, but your interests are in irrigation and you believe that orchard irrigation can be improved, then work on irrigation—that's high priority for you. (Note also that your graduate committee, funding agency, or client must share your priority interests; if they don't, you should reconsider.)

Importance of the Project

This section addresses the questions So what? and Who cares? It is a section that often is treated too lightly, or even omitted. Usually the short shrift treatment is unintentional. Students often assume that the importance of the project is obvious, or at least implied in the background and problem statements; therefore, elaboration is unnecessary. Unfortunately, readers—including funding agencies, publishers, and others in influential positions—might not innately share your insight. Spell it out for them in the research plan and in the thesis.

The section should answer the following questions:

- Why is the project needed?
- Whom will it benefit?
- How will it benefit the client?

Include specifics, and don't worry about a little overlap with points made earlier in background information. Just don't make it repetitious. Later, you will revise this part and probably incorporate it into a more general introduction of your thesis or professional paper. You can edit out redundancies at that time.

General Approach

The general approach should be given in one or two short paragraphs that outline (no details) the type of study and methods to be used. Consider the dualities presented in Chapter 2. For example, will your work be basic or applied, theoretical or empirical, analytical or predictive? If it will address both options in a duality, say so.

Next, what are you going to do? Will you chemically analyze substances, or use questionnaires to survey people? Will the study be conducted in the field or laboratory or both?

The Middle Stages

Literature Review

The literature review is like a term paper within your research plan. It is a narrative that presents the historical and current knowledge of your subject by integrating the work of numerous investigators. It brings you and your readers up to date on technical advances, including the latest methods and protocols. In addition, to conveying what is known about your topic, a good literature review will reveal gaps in that knowledge and inconsistencies in research findings. Detailed explanations for writing one are given in the section "Preparing a Literature Review," in Chapter 4.

Although the literature review normally stands alone as a separate section in your research plan, it is developed in conjunction with the plan's other components. Moreover, the literature review and the planning hierarchy feed into each other, much like a symbiotic relationship between organisms (Fig. 3-5).

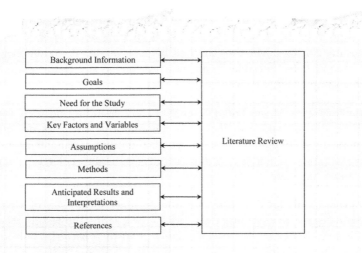

Figure 3-5. The literature review and planning hierarchy feed into each other.

Key Factors and Variables

This section will lead you to the heart of systematic inquiry. It is where you really begin to explore and exploit your thoughts and ideas, and set a definite course for your research. Every study will have key factors and variables, although they might be known by different names among various disciplines.

Key factors are the general things you need to know to solve the problem or achieve the goal. *Variables* are the specific, clearly identifiable, and usually measurable quantities used to define or evaluate each key factor. Variables are subject to change and may assume any number of forms or values, usually within a certain range. They might vary over time or change from one sample to another. Some variables are central to the research, whereas others are extraneous and unwanted but perhaps unavoidable.

You should list more key factors and variables than you intend actually to measure or evaluate. The additional items are part of the scaffolding of your plan. They help you and your reviewers understand the context of your work, they make your planning efforts more defensible, and they provide alternatives for when you run into trouble. The extras will not be needed in your final thesis; they may be deleted once they have served their purpose.

Each key factor normally has numerous variables by which it can be defined or measured. For example, in a plant productivity study, you might at first consider plant growth to be a variable, but by our definition, perhaps it is not. More probably, it would be a key factor having several possible variables, including changes over time in plant height, weight, stem diameter, and so on. Each of these variables is an indicator of the key factor: plant growth. More examples will be given later.

Variables can be quantitative or qualitative, depending on their key factor and the overall nature of the study. Quantitative variables are measured numerically, as in plant height, car speed, and number of hotel rooms. Qualitative variables are gauged nonnumerically, as in fabric color, type of protein, and emotional state.

If a project calls for detecting only the presence or absence of a parameter without regard to exact amounts, trends, or quality, the results may be left qualitative. For example, the characterization of a bacterial culture might include the following:

Key factor: carbon source metabolized
Variables: glucose (yes or no)
lactose (yes or no)
carbonate (yes or no)

In some studies, variables can be converted from one type to the other. These conversions normally require that a variable's values be considered relative to each other. For example, car speeds measured quantitatively as distance traveled per unit of time could be categorized qualitatively as slow, moderate, or fast. In a psychological study, emotional state, which is qualitative, could be assigned a numerical ranking. Although converting is not the usual case, it sometimes is helpful and appropriate for processing and evaluating data. Before converting, you should ensure that you can defend any value judgments that would be placed on the data. For example, measuring a car's speed at 40 mph is a recordable fact; categorizing this speed as moderate renders a value judgment. Likewise, ranking an emotional state from 1 to 10 might require a value judgment.

Independent vs. Dependent Key Factors and Variables

Key factors and variables may be independent or dependent. *Independent* factors and variables are those that are manipulated in an experiment or that otherwise control or appear to control the magnitude of other values. When they are part of a project's design, they often are called *design* parameters. In a graph, independent variables normally are plotted on the x-axis.

Dependent factors and variables are those that respond or appear to respond to changes in independent variables. They often are called *response* factors and variables and usually are responses that the researcher wants to identify, explain, or predict. In a graph, dependent variables normally are plotted on the y-axis. The "response" might be real or merely coincidental (see "Correlation vs. Cause-Effect," in Chapter 2).

Many studies are designed to determine how dependent variables are affected by independent variables. In cause-effect relationships, the cause is independent, and the effect is dependent. (Sometimes two variables change simultaneously, without a clear distinction between independent and dependent; these often are called *covariables*.)

A key factor and its accompanying variables are independent or dependent relative to other sets of key factors and variables. For three key factors, A, B, and C, variable B-1 might be dependent with respect to variable A-1 but independent with respect to variable C-1 (Fig. 3-6). For example, the brains of children exposed to different concentrations of environmental lead (variable A-1) might show increased activity of an unfavorable enzyme (variable B-1), which in turn correlates with a decreasing measure of brain function (variable C-1).

In your research plan, list each key factor separately, followed by its accompanying variables. Present them in outline form. (See Table 3-1 for some examples.)

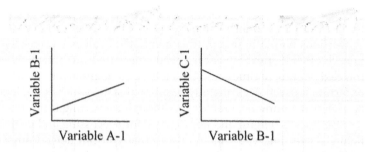

Figure 3-6. Variable B-1 is dependent with respect to variable A-1, but independent with respect to variable C-1.

TABLE 3-1. Examples of key factors and variables.

Independent key factor	Independent variables	Dependent key factor	Dependent variables
A PLANT GROWTH STUDY			
environmental conditions	temperature humidity wind speed	plant growth	height increase diameter increase root biomass increase
A WINE-MAKING STUDY			
grape quality	sugar content fruit size fruit color	wine quality	flavor bouquet acidity
A LAND-VALUATION STUDY			
location	distance from labor force distance from open water taxation zone	purchasing costs	price tax rate insurance rate
A WATER-WAVE GENERATION STUDY			
water characteristics	water temperature salt content suspended sediment load	wave characteristics	length amplitude frequency
A RECREATIONAL-PREFERENCE STUDY			
amenity characteristics	facility size admission price proximity to other amenities	user response	money spent number of users frequency of use

Controlled vs. Uncontrolled Key Factors and Variables

Variables may be *controlled* or *uncontrolled*. In the plant growth example (Table 3-1), the design variables temperature, humidity, and wind speed might be controlled, such as in a growth chamber. Any or all of them could be held at a single level or at several chosen levels. Alternatively, they could be left uncontrolled, as would occur in most field studies. The researcher must decide which approach to take. *Note that in making these decisions, the researcher effectively sets the study's scope and objective.*

Most studies will have a number of unwanted, but unavoidable, *imposed or constraint variables*, such as uncontrolled environmental conditions or genetic differences among test subjects. These must be considered and treated by one of the techniques presented later under "Choices for Handling Key Factors and Variables." Probably the best and most common approach is to ensure adequate replication and proper statistical randomization in the study design.

Aggregated or Nested Key Factors and Variables

In some studies, certain key factors and variables might be aggregated or nested. This allows them to be defined or measured at different scales. Thus a given parameter might be considered a key factor in a lower-level study and a variable in a higher-level study (i.e., lower-level variables are combined or aggregated at the higher level).

Let's consider an example of two research projects in which the first scales up to the second; that is, the research goal shifts from a more narrow focus on plant productivity to a broader view of ecosystem productivity (Table 3-2). For the first project, we'll let the key factors be plant growth and metabolism; and for the second project, we'll merge these into the more inclusive,

TABLE 3-2. Examples of segregated and aggregated key factors and variables.

Plant productivity study (segregated variables)	Ecosystem productivity study (aggregated variables)
KEY FACTOR 1: *plant growth*	KEY FACTOR 1: plant productivity
VARIABLES: plant height change plant weight change	VARIABLES: *plant growth* *plant metabolic rate*
KEY FACTOR 2: *plant metabolic rate*	
VARIABLES: photosynthetic rate respiration rate	

aggregated key factor of plant productivity. With this change in the research goal, the initially segregated variables also become aggregated.

To summarize: In scaling up the research, plant productivity drops from the goal level to the key-factor level while plant growth and metabolism drop from key factors to variables. Concurrently, the first set of variables becomes aggregated into its corresponding second set. (Although we recognize that ecosystem productivity is a function of all biological activity, we have simplified this example to include only the plant contribution. Also for simplicity, independent and dependent variables are not distinguished.)

Note that when the goal focuses on plant productivity (the first case), the variables most probably are measured specifically and directly for the study. But under the goal of ecosystem modeling, data might be collated from different sources and plugged into the model. Data from different sources might have been measured in different ways. The ecosystem researcher then would have to assume that the data are appropriate for the current work. This is a good example of where the dualities theoretical vs. operational definitions and fact vs. assumption apply.

Now let's look at an example in which focus shifts from broader to more specific levels. This time we'll consider research into business production costs (Table 3-3). (Again, for simplicity, independent and dependent parameters are not designated.)

From this set of examples, a study concerning company-wide conditions or trends would encompass the broadest key factors and assess aggregated data as variables. Alternatively, a study of an individual cost category would focus on a more specific level with variables consisting of less aggregated data.

Finally, note that in these examples we have used a hierarchal approach to guide us in generating, sorting, and evaluating goals and potential parameters. Moving alternately up and down the hierarchy helps us to see and assess relationships that we might otherwise overlook or misconstrue. Moving

TABLE 3-3. Examples of aggregated and segregated key factors and variables.

Broad category	More specific category	Most specific category
KEY FACTOR:	KEY FACTOR:	KEY FACTOR:
company production cost	*worker cost*	*worker benefits cost*
VARIABLES:	VARIABLES:	VARIABLES:
worker cost	salary or wages	worker's compensation
factory operation cost	*benefits*	vacations
transportation cost	bonuses	health care

within the hierarchy prompts us to reexamine and possibly revise earlier visions and decisions. In this way, we exercise the system of checks and balances that is inherent in this type of approach.

Choices for Handling Key Factors and Variables

Variables can be handled in any of several ways depending on the purpose of the work:

1. They can be integrated into the study. They can be manipulated, as might be design variables, or measured and interpreted, as with response variables.

2. They can be standardized or eliminated. For example, a parameter (e.g., temperature, price, etc.) might be held constant, or potential parameters could be excluded (e.g., limit the study to a single gender of human subjects or a single metal alloy).

3. They can be statistically randomized. If done with proper design and adequate replications, this approach effectively minimizes their impact on the interpretation of results.

4. Their influence or response can be assumed. This option should be cautiously evaluated and supported by evidence in the literature review.

5. When all else fails, they can be disregarded. This choice may be difficult to defend, but nonetheless it sometimes is unavoidable. It should be acknowledged and explained.

Whichever approach you take, give your key factors and variables careful thought. Address them in your literature review, and discuss them with your graduate committee. They are perhaps the most important part of your research plan, because without them you will not be able to define the scope of your work, develop a proper objective, or select appropriate methods.

Scope

The scope sets the boundaries or limits of your research. It should be presented in outline form and may be defined by time, space, populations, key factors and variables, equipment, or any other limitations you select. The following are examples of items that might be included in the scope of various kinds of projects:

Time

- limited to six months of data collection

- limited to one complete growing season

- limited to the Christmas shopping season

- terminated when equipment has burned 500 liters of fuel.

Geographic Location

- limited to a particular city or county

- limited to the distribution of a particular geologic formation

- limited to a particular experimental forest, farm, or research station

- limited to the dark side of the Moon

Populations

- limited to specified bacterial strains

- limited to volunteer subjects from the state prison

- limited to subjects showing symptoms of vitamin deficiency

- limited to dairy cows on the campus farm

Variables

- limited to 5-, 10-, and 20-ohm resistors

- limited to differences in fruit yield

- limited to treatment by 0.5 M HCl at room temperature

- limited to evaluation of thiamine and riboflavin in the human diet

Sometimes, in addition to listing what the project will include, the scope should mention what it will not include. Obviously, this list could extend indefinitely, but the point is to acknowledge seemingly relevant factors that you have decided not to investigate. Mentioning them at this stage of your work shows your reviewers that you are not simply overlooking important obvious points but that you have considered and rejected them. This technique helps put you in a defensible position later—at your oral defense for example.

Objectives and Hypotheses

The objectives and hypotheses section is placed midway through this chapter to emphasize that preceding sections must be addressed before objectives and hypotheses can be adequately defined or formulated. The literature must be reviewed, key factors and variables selected, and the scope of the project laid out before proper and final objectives can be written.

Despite all the preliminary work, you will begin to visualize objectives and entertain hypotheses long before you reach this point. Do not delay writing drafts of them; state them as best you can as soon as your thesis topic is selected. *The objective is the most important part of your research plan—it should be written early in the process and revised until you get it right.*

An *objective* (or "object," as some people prefer) should be a clear, straightforward statement that answers the question, What, *specifically*, will the project or a particular part of the project accomplish? In the planning hierarchy (Fig. 3-4), the overall goal branches to subgoals, and a single subgoal branches through multiple key factors and variables objectives. Depending on the selection of key factors and variables, a study may be limited to a single objective, or it may address several related objectives.

Good objectives are more than exploratory; they are designed to close gaps and expand knowledge. Try to express each objective such that it leads not only to methods and anticipated results, but also to criteria for judging the validity, soundness, and reliability of results. These criteria need not be stated in the objective, but they should follow logically from it. For example, if your work is quantitative, your objective might be framed to support a statistical design and analyses. In the end, you should be able to state whether or not a given objective has been accomplished.

In addition to objective statements (or sometimes perhaps in lieu of them), hypotheses should guide the study. As explained in Chapter 2, hypotheses are tentative and testable explanations of conditions or phenomena (see "Deduction vs. Induction" and "A Triad: The Testable, the Tested, and the True"). (Note: Statistical testing normally requires a null hypothesis and an alternative hypothesis. In effect, the null hypothesis states that any two variables are related only by chance. In statistical hypothesis testing, emphasis is on the null hypothesis, which we either reject or fail to reject. Consult a statistics text for details.)

Here are some criteria for creating useful hypotheses:

- Keep them simple and specific. If you have multiple points to test, place each in a separate hypothesis, allowing for individualized testing and revision.

- Delineate them clearly enough to set a course, but don't cut them so deeply that they discourage consideration of alternate routes.

- When you have two competing hypotheses, chose the simpler one first. (Choosing the simpler one is known as the principle of parsimony or Ockham's razor, which says cut out the complications. The principle was advocated by William of Ockham (or Occam), a 14th-century English philosopher.)

Be careful not to get too comfortable with your hypotheses. Of course you won't be able to view them dispassionately—if you could, you'd feel no excitement in coming up with a bright idea, and lacking excitement, you'd have to be emotionally inert—nonetheless you should prepare yourself to abandon a hypothesis when it leads nowhere.

Consider multiple, competing hypotheses, so that you readily can change direction or pursue an alternate route if necessary. This point hardly can be overemphasized. Error in hypotheses is a normal part of the scientific research process—it should not be considered failure. Remember: *Your hypotheses should be tools, not your personal desires. Your reward should be in figuring out how something works, not in proving that your favorite idea is correct.* Still, no matter how hard you try, times will come when rejecting a hypothesis will require you to refocus your ego. (Refer to "Science vs. Advocacy," in Chapter 2.)

Lastly, bear in mind that you can change your objectives at any time, even after you've worked out your methods and begun collecting data. Despite your best planning efforts, sooner or later you could find a discrepancy between what you said you would do and what you actually are doing. If you discover this, don't immediately conclude that you've been collecting improper data. Perhaps your methods are fine. The problem could lie in an improperly defined objective. Use your vertical intellectual slider to figure it out, and if you find that your objective is defective, fix it.

Place your final objective statement(s) near the beginning of your research plan and thesis. Hypotheses may be placed with the objectives, or they may be included later in the methods section.

Assumptions

Assumptions are factors, variables, conditions, or anticipated events that are taken for granted or supposed (see "Fact vs. Assumption," in Chapter 2). They are uncontrolled, unproved, and too often unchallenged; nonetheless, we tend to be confident that their state or outcome is, or will be, favorable. At times, this confidence is so strong that the assumptions are taken as fact, considered uninfluential, or overlooked entirely.

Although they are taken for granted, not all assumptions are innocuous— some are malignant. A malignant assumption that is not properly considered can invalidate a study or prompt an interpretation of results that later proves embarrassing. Therefore, you should try to identify all significant assumptions on which your study is based. Some assumptions will be easily recognized and evaluated. Others will be obscure and difficult. Some studies will lack any significant assumptions.

Assumptions are common in economic and marketing studies. Economic predictions, including those of consumer motivation and preference, and nearly all applications of price theory, rest on assumptions. Plant growth and yield studies sometimes rely on assumptions such as adequate water or nutrient availability and favorable or even uniform environmental conditions. A human nutrition study might depend on the cooperation of test subjects to report their health, eating habits, and life style. Often the investigator must assume that these reports are accurate and reliable.

Methods and Experiments

This section of your plan should be a detailed account of methods used to carry out your project. It is where more universal research planning diverges onto the more discipline-specific tracks of research methods or research design (refer back to comments in the preface).

Because of differences in methods and protocols among various sciences, the items to include will vary greatly among disciplines and types of studies. Seek advice from your graduate committee, and refer to theses of other students and to journal articles in your field for content and formatting ideas. Check your university library for research methods books that pertain to your area. In addition, you should consider taking an experimental design and analysis course that includes statistical procedures.

No matter what your field, your research methods should strive to do the following:

- Frame your research with sideboards and endpoints that restrict you from wandering too far into extraneous (albeit interesting) fields and that stop you when you've gone far enough.

- Control inductive extremes by limiting the number and type of observations, and control deductive obsessions by doing the same with hypotheses.

- Provide mechanisms for recognizing and choosing among competing possibilities.

- Provide for clear and frequent feedback into and from other parts of your research plan, especially key factors and variables, scope, and objectives.

- Minimize chances of bias—both intellectual and sensory.

- Provide for critical analysis as well as data generation.

- Define steps clearly but allow for flexibility and innovation.

- Allow for imaginative thought as well as critical thought.

- Distinguish processes of validation from those of falsification, and allow for repair of hypotheses.

- In hypothesis-driven research, design the study to falsify hypotheses.

- Provide mechanisms that ensure proper discrimination between discovery and justification of results.

- Provide enough detail to direct you efficiently along your research path and to allow critical review of your progress.

You've probably noticed that most of the points listed above are repeated from earlier sections of the research plan, and others were discussed in Chapter 2. This redundancy is intentional—it's part of the system of checks and balances that should run throughout research planning. It's also a reminder for you to operate your intellectual slider vertically, up and down the planning hierarchy. Check your methods against other parts of your plan. Do they properly address key factors and variables listed in the scope? Are they compatible with your objectives and, looking ahead, with your anticipated results? Used in this way, your slider becomes the instrument for maintaining checks and balances and ensuring consistency in your work.

Your study might or might not involve experiments. Lacking experiments, your work could take any of several approaches, including building a theoretical model that relies on existing data or characterizing a system without manipulating its components. In either case, your efforts should yield new information, and you should understand what distinguishes experimentation from other activities, including demonstration.

Whereas experiments lead to new knowledge through hypothesis testing, demonstrations reiterate what we already know. Often, "experiments" conducted in high school and in many undergraduate courses are merely demonstrations of the Aristotelian type—that is activities designed to illustrate preconceived concepts and convince students of their validity (Medawar, 1969).

A second level of activity, which begins to link demonstration and experimentation, relies on simple Baconian induction or contrivances. By this method, the investigator does something and observes what happens or gathers data to see what they might yield. Although this technique approaches scientific research, it falls short because it lacks critical analysis.

Moving up another level, we come to Galilean experiments, which set and test hypotheses designed to differentiate and select between opposable options. A great deal of successful research conforms to this model.

Fourth, we reach Kantian experiments, which can transcend the Galilean model by inducing us to question and alter the very premises on which our hypothetico-deductive scheme is based. Designing experiments of this caliber requires us to recognize that our senses—or at least the way in which we interpret sensual stimuli—might not be reliable, even though our receptors suggest that they are. We therefore must strive to view our world differently by reaching beyond conditioned biases of our senses and emotions (i.e., beyond "common sense") (Kant, 1900; Medawar, 1969). This can be an exceedingly difficult endeavor, requiring years of experience, practice, and concentration. Few graduate programs would demand that students master this level, nevertheless the good ones would expect them—especially doctoral students—to seriously try.

Finally, we might ask what makes a good experiment? Medawar (1969) offered an elegant response: "a 'good' experiment is precisely that which spares us the exertion of thinking: the better it is, the less we have to worry about its interpretation" (p. 15). Beautifully stated and catchy, but in this context, Medawar neglected to emphasize that the spared exertion is enjoyed on the outcome side of an experiment. In formulating good experiments, we are spared no exertion on the design side.

Apart from pondering the foregoing considerations about experiments, writing your proposed methods should be fairly straightforward. Here's a checklist of points you might need to include:

- location and description of study sites
- time and duration of the study
- hypotheses to be tested

- population sampled and sample size

- sampling scheme or protocols

- sample selection, storage, and preparation procedures

- scientific names of organisms

- experimental and statistical designs

- characteristics of test subjects

- development of questionnaires

- treatment descriptions

- theoretical and operational definitions (see Chapter 2);

- analytical methods

- instruments and equipment used

- model development, calculation procedures or formulas

- specialized computer programs

- equipment construction

- data analysis or techniques for interpreting results, including statistical tests

- falsification or validation criteria

- quality control criteria and procedures

Standard procedures described in methods-of-analysis books will not require detailed explanations, but they must be properly cited and referenced. Explain ways in which your methods deviate from published procedures. Use flow charts, diagrams, or figures to clarify complex procedures and tables to present materials lists. Use generic rather than trade names whenever possible. Again, be sure to cite references.

Begin keeping a proper research notebook (see Chapter 5), in which you record your detailed methods and progress. The notebook will help ensure accuracy and greatly ease the writing of your final methods section. But perhaps most important, it will help you maintain a clear and defensible position throughout.

When you finish your research, your methods write-up should allow other investigators to reproduce and verify—or in some cases question—your work. Having at least some aspect of your work questioned should not be feared—every good scientist goes through it, and without it science cannot progress. Remember that in open, honest science, challenges should be issued to strengthen mutual benefit, not to demonstrate arrogant superiority. Still, being challenged on technique or interpretations is one thing; being challenged because of improper documentation or sloppiness is quite another.

The Closing Stages

Anticipated Results

This section of your research plan will present predicted results—even before you've worked out a theoretical model or collected any data. Writing this section early on will allow you to test the soundness of your objective and overall project design before committing costly time and resources to a possibly flawed study. Your results should be compatible with all that goes before—methods, objective, scope, key factors and variables, subgoal, and overall goal. If they're not compatible, you have a problem. The good news is you'll know it, preferably before anyone else does.

Writing your results in advance also allows you to reveal your expectations and personal biases. Personal biases are OK. We all have them. They're not necessarily bad or scientifically invalid, provided you don't drift from science to advocacy (see Chapter 2). The best approach is to recognize biases and present them forthrightly, rather than trying to repress or hide them. Biases can be very helpful in developing hypotheses. And like hypotheses, they should be tested.

Begin by laying out the key factors and variables to be included in your study. Make up some "dummy" data, and illustrate your predicted findings in properly constructed tables, graphs, charts, or other illustrations that might be included in your thesis. In other words, set the table before you serve the data. (Later, you might even practice the appropriate statistical analyses on selected data.)

After constructing the tables and figures, back up and begin the results section by writing a brief narrative—a declarative statement—of your anticipated results. Refer to tables and figures parenthetically in the narrative, rather than making them the subject of a sentence. In other words, avoid writing "Table 1 shows . . ." or "The data are presented in Table 1." Also avoid this empty phrase: "The results show that . . ." These types of openings are merely "wheel spinning." They lay down words without going anywhere. More importantly, they fail to focus your thinking on where it should be, which is the message of the table's data, not the table itself (see "Hints for Scientific Writing," in Chapter 4).

Set the table before you serve the data.

Keep tables, figures, and other illustrations in proper sequence and numbered accordingly. Place each one after its first narrative reference, either on the same page or the following page.

Here's an example of how you might start your narrative: "Soil nitrogen concentrations probably will decrease significantly, and pine needles will turn yellow within three months after herbicides are applied to the soil (Table 1)." (See Fig. 3-7 for the Table 1 referred to in this narrative example.)

In the example table (Fig. 3-7), note that both design and response variables are presented, as are independent and dependent variables. Recall that design variables, which are set by the researcher, are independent. Response variables are dependent with respect to design variables, but they can be either independent or dependent with respect to each other.

In this example, design variables include herbicide application rates and measurement times. The response variables, each of which measures a different key factor, are soil nitrogen concentration and needle color. Soil nitrogen is dependent with respect to herbicide application rate and timing, but independent

Table 1. Soil nitrogen (N) and pine needle color responses to atrazine applications.

Time after application	Soil N			Needle color		
d	g kg⁻¹ Replication			Munsell notation Replication		
	1	2	3	1	2	3
0	5.1	5.0	4.8	5G 8/6	5G 8/6	5G 8/6
30	5.0	4.9	4.6	5G 8/6	5G 8/6	5G 8/6
60	4.5	4.4	4.7	5G 8/6	5G 8/6	5G 8/6
90	4.1	4.0	3.9	5G 8/6	5G 8/6	5G 8/6
0	5.2	4.9	5.1	5G 8/6	5G 8/6	5G 8/6
30	4.3	4.0	4.0	5G 8/6	5G 8/6	5G 8/6
60	2.6	2.2	2.3	5GY6/4	5YG6/4	5GY5/4
90	1.1	0.9	1.2	5GY6/2	5YG6/2	5GY5/2
0	5.0	5.3	5.3	5G 8/6	5G 8/6	5G 8/6
30	2.8	2.4	2.1	5GY8/4	5GY8/4	5YG8/4
60	1.3	1.2	0.9	5GY8/2	5GY8/2	5YG8/2
90	0.6	0.3	0.4	2.5Y6/1	2.5Y6/1	2.5Y6/1

Figure 3-7. Example of a properly constructed table. (Note: These data are not real; they are for illustration only. A better study also would relate foliar N to needle color and soil N.)

with respect to needle color. Needle color is dependent with respect to all other variables.

Although needle colors are dependent with respect to soil N, these data do not show that soil N controls needle color. A correlation is suggested, but a cause-effect relationship is not proven (see "Correlation vs. Cause-Effect," in Chapter 2).

Once you've commenced your actual study, you might produce copious amounts data, enough to fill several pages of tables. Should you present all of this in the results section of your thesis? Probably not, but check with your committee first. Usually lengthy data, especially with numerous replications or nonessential details, may be placed in an appendix. Make up a summary table or figure to place in the results section.

Figure 3-8 illustrates a properly constructed figure. Note that it is two-dimensional. A three-dimensional figure would not be appropriate. Use three-dimensional figures only to present data that require x, y, and z axes in a single chart.

Because this manual focuses on research planning rather than on research methods, it presents only a simple example of a table and a figure. Many of

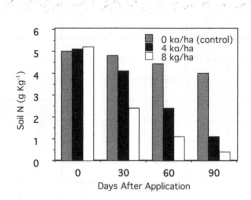

Figure 3-8. Response of soil N to various herbicide application rates during the first 90 days after application.

you will generate data that require more complex and sophisticated forms of presentation. Numerous excellent examples of tabular and graphical displays can be found in the literature (e.g., Valiela, 2009), and numerous computer programs are available to help you generate almost any type of data display.

Finally, ensure that your anticipated results are congruent with your objective and all other parts of your research plan. If you find inconsistencies, now is the time to correct them. If you have trouble with anticipated results, you probably have a fuzzy objective or improperly defined key factors and variables. (A good exercise at this time is to write an objective statement for the information presented in the table in Figure 3-7. Try it.)

Citations and References

Citations are placed as needed throughout the body of your thesis or professional paper. They identify the authoritative sources (i.e., references) of information gathered from professional journals, textbooks, the internet, and other sources, including personal communications. They give only the author(s) and year (e.g., Strunk and White, 2000; Taskey et al., 1989) of each reference. (Use "et al.," the abbreviation for the Latin *et alia*, 'and others,' for three or more authors.) Alternatively, they can give numbers that are keyed to the references. Enclose citations in parentheses immediately after the pertinent information. Make sure that each citation corresponds to a full reference listed in "References."

(Note: Use the author-year format for citations in your research plan and thesis. It's easier to track and update than the number system. Journals often require the number format to save space, so be sure to check. Switching from author-year to numbers is easy.)

References are placed together at the end of your research plan, thesis, and professional paper. They give your reader all the information needed to find each source. Include the names of all authors ("et al." must not be used in references), year of publication, title of article, name of journal or other source, volume number, pages, and place of publication. The exact information and format required vary among graduate programs, professional organizations, and other publishers. Ask your advisor for guidance, and take a look at a few germane professional journals.

Most large professional societies publish instructions (in book form, on the internet, or both) for preparing publishable manuscripts. These include rules for formatting citations and references. The formats tend to vary only slightly among most of the physical and natural science societies, but these are significantly different from the formats used in the social sciences. Check the one accepted in your field. Two professional societies that publish exceptionally helpful style guides are the American Society of Agronomy (ASA), which puts out *Publications Handbook and Style Manual* (ASA, 2011), and the American Chemical Society (ACS), which makes two editions of its style guide available: *The ACS Style Guide: A Manual for Authors and Editors* (Dodd, 1997) and *The ACS Style Guide: Effective Communication of Scientific Information* (Coghill and Garson, 2006). (The two ACS editions differ significantly, with the later edition focusing more on electronic publication; both are recommended.)

Once you know the requirements for citing and referencing your literature sources, you can increase efficiency and decrease errors by using appropriate software to format and collate your references. In the meantime, the formatting used in the reference section of this book will get you started.

Publication Plans

This section merely states the means intended for reporting your work. For some of you, the report will be limited to the thesis submitted to your university. But for many others, especially those pursuing a doctorate, plans should include one or more articles in a refereed scientific journal. Name the journal(s) and give the number of articles you hope to publish. If your program requires journal publication, you should discuss this with your graduate committee early in your studies, then plan accordingly.

Timetable

Present a calendar or flowchart of activities and events. Include the estimated time required for major activities and anticipated milestone dates. For some kinds of studies, a critical path analysis might be appropriate. Your word processor or another computer program can help with timetable construction. When you think you've figured out the amount of time needed for your final write-up—triple it, unless you've been diligent about keeping up with the writing for your research plan. Even then, you'll probably underestimate.

Budget

Every project, regardless of its size or success, costs money and consumes resources. Some projects are small—so small that the student might get the impression that they don't cost anything; nonetheless, someone pays the bill even if a project utilizes existing equipment, faculty receive their normal salary, and the student works for nothing.

Your research plan should have a budget, even if your project is unfunded. Usually the chief budgetary responsibility falls to the major professor, who figures out the costs and writes a grant proposal, often in collaboration with the student and perhaps other professors. Most universities have an affiliated foundation or a sponsored-programs office that processes grant proposals and coordinates with granting agencies. The professor works through the appropriate foundation or office. Ask your advisor or committee members to inform you about the budget and to tutor you in budgetary matters.

Sometimes, in absence of a designated grant, the costs for an individual student are absorbed into existing budgets. In these cases, the student often bears part of the cost, usually by foregoing monetary support. This situation occurs especially at the master's degree level in institutions having small research programs. No matter how they are covered, costs still are incurred and you should account for them.

Normally, the budget is presented in a table (Fig. 3-9), but unusual items may require additional explanations. Also, the method of calculating certain costs should be made clear. For example, the basis for salary amounts and travel should be specified (e.g., $X/h \times Y$ h = Z; X miles $\times$ Y/mile = Z), as should the rates for staff benefits and indirect costs.

The following checklist explains common budget items:

   1. **Salaries and wages.** Consider yourself and all other university
      employees directly involved in the work, including the principal
      investigator (PI) (usually your major professor) and other faculty,
      post-doctoral associates, technicians, undergraduate assistants,
      secretaries, and on-campus consultants. Specify the amount
      of time or proportion of their appointment each person will

Budget: Insert project title.
Insert project beginning and completion dates.

	Amount (US $)		
	Sponsor	In-kind	Total
1. Salaries and wages			
A. Faculty:			
Marie Curie (PI) (12 mo @ 40%)	25,000	15,000	40,000
Charles Darwin (10 wk @ 100%)	16,350		16,350
B. Graduate Students:			
René Descartes (12 mo stipend @ 100%)	28,950		28,950
Rosalyn Franklin (9 mo stipend @ 100%)	21,600		21,600
C. Technician:			
Arthur Casagrande (640 h @ $18/h)	5,760	5,760	11,520
2. Fringe benefits			
A. Faculty: 25.5%	10,544	3,825	14,369
B. Graduate Students: 18.6%	9,402		9,402
C. Technician: 22.7%	1,308	1,308	2,616
3. Equipment			
Atomic absorption spectrophotometer	62,000		62,000
Quantitative polymerase chain reaction		12,700	12,700
4. Supplies			
Laboratory glassware		1,800	1,800
Test plants		450	450
5. Travel			
SPCGS Conference (PI and graduate student)	5,750		5,750
Sample collection		750	750
6. Services			
Statistician		3,800	3,800
7. Indirect costs			
50 % of MTDC (see budget checklist item 8)	32,356		32,356
8. Cost sharing			
X-ray diffractometer (Grant TU-NSF-08-03)			67,800
9. Totals	219,020	45,393	332,213

Figure 3-9. Example format for a research budget.

contribute to the project. Include anticipated pay raises for projects that will run several years.

2. **Fringe benefits.** This category covers medical insurance, vacation time, retirement accounts, and related allowances. Amounts vary for different positions and pay scales but typically range between about 20 and 30% of the salary. Sometimes formulas vary with the type of project and funding source.

3. **Equipment.** Commonly, a durable item costing more than a certain amount (e.g., $5,000), including installation, is considered equipment. The definition varies among granting agencies and changes with time. Equipment rental should be included if needed.

4. **Materials and supplies.** Include office supplies, telephone, duplications, test animals and their food, and laboratory supplies and chemicals. Do not list items separately; lump them into categories.

5. **Travel.** Figure a cost per distance or rental rate for automobiles, vans, and trucks. Normally, the rate for vans, trucks, and four-wheel-drive vehicles is greater than that for automobiles. Some studies might require routine airplane or boat travel or rental. Include travel, lodging, and per diem (i.e., daily cost for meals) for field work, consultation, and professional meetings. Consult your major professor, university foundation office, or a granting agency for current rates.

6. **Services.** Consider costs of computer time, large duplication services not included in supplies, photographic services, and publication. (Publication won't be free—you'll have to pay for it.) Some studies might require special legal reviews, as when proprietary rights or human subjects are involved. Still others must include costs of hazardous waste disposal.

7. **Other.** Perhaps you will need special health insurance, space rental, books or journal subscriptions, or academic fees and tuition.

8. **Indirect costs (overhead).** This amount reimburses the university for use of the facilities and administrative expenses. Among other things, it pays the light and phone bills, internet connections, janitor, library, mortgage on the laboratory, and administrative help in securing and managing the grant. Securing and managing the grant can be a big deal because numerous federal, state, and university regulations must be met, and proper cost-accounting standards must be followed. For these, you need expert advice and, in some cases, legal protection.

The rate, which usually is negotiated between the university or university foundation and the granting agency, is highly variable. For U.S. federal government grants, it might run about half of total salaries and wages, or from about a quarter to two-thirds of total direct costs or modified total direct costs (MTDC). (Modified total direct costs exclude equipment, scholarships, and certain off-campus activities, but again definitions and exclusions vary.) Charitable foundations normally negotiate a lower rate, varying from zero to about 25% of total direct costs. Sometimes, a lower rate applies if the bulk of the research is conducted off campus. Be sure to check with your university's grants development office for the proper rate and method of calculation.

After figuring all your anticipated costs from items 1–6 above, you must add indirect costs to the total. Remember, if the rate is based on a percentage of wages and salaries, and midway through the project you decide to shift funds into wages and salaries from another category (e.g., from travel), your project likely will be charged the additional overhead. Result: you have less money to spend.

9. **Cost-sharing or "in-kind contributions."** When included, these show funding or other support that will be provided by sources other than the sponsor to whom the proposal is directed. Cost-sharing can be money from a separate existing grant, a pending grant not yet secured, or a university endowment. The amount of cost-sharing normally is summarized as a line item. In-kind contributions typically consist of noncash donations or loans of goods or services from the university or individuals. Examples include equipment on hand and some proportion of faculty time. These usually are itemized in a separate column of the budget. Cost sharing and in-kind contributions can be especially important when matching funds are required to secure a grant. As with indirect costs, they require expert help to negotiate and administer.

Appendices

Appendices might include the following:

- glossary of terms
- data sheets
- questionnaires
- detailed materials, supplies, and equipment list

Research Grant Proposals

A research grant proposal is a request for funding. It attempts to persuade someone who controls money that his or her organization should pay for your envisioned project, or at least part of it. To be successful, the proposal must hook reviewers' attention and convince them that it fits their organizational mission and funding criteria. Reviewers must feel assured that the study will be conducted competently and, in most cases, that it has a high probability of success.

The proposal, which normally is written before the research plan, sketches out the research ideas and addresses many of the same topics, but being preliminary to the plan, lacks its full development. Typically, a proposal briefly describes the research problem and goals, accentuates the study's significance, highlights past and current progress, outlines methods, and points up the investigators' credentials and competence. It focuses on the budget and time line, and demonstrates how the money will be well spent.

Research grants can be secured from a multitude of sources, including government agencies (federal, state, and even local), philanthropic and science foundations, businesses and corporations, and private benefactors. Although huge amounts of money are awarded each year, competition for funds can be fierce. For this reason, most researchers submit multiple proposals annually to keep their research facilities running.

Numerous books and internet sites provide excellent instructions for writing grant proposals, and many granting agencies publish detailed guidelines for funding requests. In addition, research universities and an army of private consultants offer courses and workshops that can teach you how to locate and secure funding. Check online, and you will find computer software designed specifically to help you with the job.

Given the wealth of easily accessible information and the variety of requirements among granting organizations, this manual need not elaborate on the subject. The main purpose here is to make you aware that money and instructions for obtaining it are available—you don't have to go it alone. So log into your favorite internet search engine, type in "research grants" or comparable wording, and you're on your way. Be sure to discuss the subject with your major professor and visit your campus research and grants development office.

Many of the points required in a research proposal are similar to those in a research plan, but the organization, formatting, and degree of detail might differ significantly, as could definitions of some terms. But if you're familiar with the guidelines for research planning presented in this chapter, making the transition to research grant proposals should be fairly easy. Here are a few things to keep in mind:

- The first rule is to follow all instructions precisely and meet the deadlines.
- Make clear the study's rationale and relevance, including its social and economic benefits.
- Choose your wording to fit that of the organization.
- Explain any preliminary work you might have done and include a data summary.

- Identify research team members and their positions. Include a brief statement of qualifications for each.

- Reveal current sources of support, if any.

- Point out any in-kind contributions in the budget.

- Submit the proposal properly, the way they want it.

- The last rule is to follow all instructions precisely and meet the deadlines.

CHAPTER 4

Reading and Writing

What's This Chapter About?

Reading the scientific literature

Writing for clarity

Writing a scientific literature review

Writing for Scientific Publication

What's In This Chapter?

Reading and Evaluating Research Reports

Previewing
Evaluating

Hints for Scientific Writing

Paragraphs
Sentences
Punctuation
Word Choice
Variety

Free Writing

Preparing a Literature Review

What Is a Literature Review?
Why Write a Literature Review, and What Should It Accomplish?
When Should the Literature Review Be Written, and How Long
    Should It Be?
What Topics Should Be Included?
What Are the Various Types of References, and Where Are
    They Found?
How Does One Organize and Prepare a Literature Review?
What about the Writing Style?
Establishing Credibility
What Is the Proper Format for Citations and References?
Avoiding Plagiarism

Writing for Scientific Publication

The Process
Get Ready
Write the Paper

Words and eggs must be handled with care.

Anne Sexton, *Words*

You already know how to read and write, but to succeed as a graduate student, you must refine these skills and exercise them at a continually high and demanding level. Reading and writing, together with thinking, are your most fundamental and important activities. You will spend your entire graduate experience trying to improve them—and with concentrated effort, you can do it.

Reading and Evaluating Scientific Reports

As a graduate student, you will spend a great deal of time reading books, monographs, and especially scientific papers (i.e., original research reports). Your most intensive reading should be during your first year as a master's-degree student, and your first two to three years as a doctoral student. You must strengthen your fundamental knowledge in your field, come up to date on the latest developments and trends, and learn who's doing what and where. You'll also need to relate your own work to that of others. Once you're heavily into your research, you'll still need to keep up with the field, but you won't have as much time for reading scientific papers, so you must build your foundation early. Learning to read critically is one of your most important tasks as a graduate student.

Previewing

Scientific papers can be difficult reading. For one reason, the subject matter might not be familiar to you. The terminology might be strange, and formulas and equations might be confusing. For another reason, some scientists simply do not write clearly and concisely. Although reading a five-page technical paper will be far more time consuming than reading five pages of a novel (as it *should* be), you can reduce your time and increase your understanding and retention by previewing the paper before reading it all the way through. Here's how:

- First, quickly determine whether or not the paper is worth your time—if it's not, move on. Read the title, scan the abstract, and glance through the pages. Often, you might find only a single page or figure that suits your purpose. You don't have to read the whole paper. But if your advisor recommended it, read it. If the whole paper looks worth reading, the next steps can help.

- If you're reading a hard copy of the paper, have a pencil in hand to take notes. Do not mark directly on the paper unless it is your personal copy. (Libraries pay big money for scientific journals.) If you're reading the paper on your computer screen, have a pencil and note cards handy anyway.

- After glancing through the paper to see its length and overall structure, think about the title. Do you understand all the words? If not, look them up now.

- Read all of the division headings and captions of tables and figures. After reading each caption, look over the table or figure, noting the variables included and some typical values. Decide which, if any, of the tables and figures contain information that might be useful to you.

- Go back to the first page, and read the abstract. Don't worry if you don't understand all of it at this time, just read it.

- Read the first paragraph and the first sentence of each succeeding paragraph. Jot short notes about each relevant table or figure as you do this.

- Now read the paper through and take notes, preferably on note cards. Write in your own words, but if you feel that you must copy a phrase or sentence word for word, be sure to place it between quotation marks; you can alter the wording to fit your own style and context later. Look up the meaning of words you don't understand. (For more suggestions on note taking, see "Preparing a Literature Review," later in this chapter.)

Previewing might seem time consuming at first, but stick with it long enough to give it a fair chance. Your speed and comprehension will increase. With experience, you can modify the process to fit your personal needs and working style.

After a while, you'll learn that you don't need to read every word in many scientific papers. You might read one paragraph thoroughly, skim another, and merely glance at the headings and a few words of others. Whether you read a paper word for word or limit yourself only to relevant points might depend as much on your personality as on the paper's content. Either way, you will be well served in your professional career if early in your academic career you learn to read, analyze, and absorb meticulously. Once you acquire the skills, you can choose when and how to apply them.

Evaluating

As a graduate student, you must develop analytical skills needed to interpret and communicate scientific information presented in a variety of publications and formats. These skills can be developed through many activities and exercises, including critically evaluating the works of others (Kuyper, 1991; Leedy, 1974; Leedy and Ormrod, 2005). The following checklist will help you critically read and critique scientific papers.

- Does the title clearly and accurately represent the paper's content?

- Is the overall problem to which the research relates clearly stated?

- Is sufficient background information (including literature review) presented to set the stage and to inform the reader of the need for the study and its application to related work?

- Are the specific objectives of the research made clear?

- Does the research appear to have been well planned and organized?

- Is previously unpublished information presented, and if so is it relevant?

- Are any hypotheses stated?

- Are the hypotheses and objectives adequately designed to achieve the research goals?

- Are field, laboratory, and analytical methods clearly stated?

- Are the methods appropriate, or can you suggest better methods?

- Are the tables and figures clearly presented, and can they stand on their own?

- Does the paper clearly indicate whether or not the hypotheses have been supported or rejected?

- Were the data interpreted?

- Are the interpretations and conclusions justified by the facts presented?

- Is the writing clear, concise, and grammatically correct?

- Can you identify any of the dualities presented in Chapter 2 of this book?

- What can you learn from this paper to help your own work?

- If you were the editor of the journal, what comments would you give to the authors?

As you study a paper, be aware that not everything in it was designed and accomplished as neatly and timely as reported. Research science can be messy— at times having parts that are more patchwork than hierarchical. Although research must be planned, some of it advances as an interrupted succession of haphazard events that include chance or otherwise unforeseen factors.

Because of the sometimes patchwork nature of research, the reasoning behind the work might not be accurately represented (Medawar, 1967; Ladd, 1987; Kuhn, 1996). A good deal of what passes for deduction, for example, probably came after the fact. That is, someone's imagination stimulated an idea, it was developed inductively, then applied deductively and reported as if it had been envisioned that way all along.

In a literal sense, even reputable scientists routinely misrepresent and distort the details of their work—rarely to commit fraud (although unfortunately that does happen) but commonly to streamline the telling of a tortuous journey. Research reports rarely give an accurate impression of the time and effort invested in the work. An idea that took months to formulate comes across in only a few lines, and mistakes that prompted reenactment of an experiment are not reported. Scientific journals do not allow space to elaborate on all the wrong turns and detours. These and other types of historical distortions magnify with the telling: they are rife in secondary and tertiary references, especially textbooks.

Hints for Scientific Writing

You can read, observe, think, and collect data, but to bring it all together, you must write—and write frequently, sometimes with ease and pleasure but mostly with strength of purpose. As a learning and research tool, writing transcends simple reporting. It can stimulate your imagination and illuminate hidden connections. *In general, if you don't understand something, you can't write it. Conversely, if you can't write something, you probably don't understand it.*

How can you improve your writing? Like anything else, the first step is to learn and practice the fundamentals. If you are rusty on these, back up and relearn basic grammar, sentence structures, and paragraphing. Spend a couple of hours studying an old-fashioned high school writing handbook—you probably can pick up one cheap in a used book store. You'll also find *The Elements of Style* by Strunk and White (2000) to be an indispensable little volume, but if you want a more in-depth authority look into *The Chicago Manual of Style* (University of Chicago Press, 2010). For well-founded guidelines and practical advice on preparing and publishing your research paper, check out *Getting Published in the Life Sciences* (Gladon et al., 2011). And of course you'll find many hints—including lots of bad ones—on the internet. Check out how to diagram sentences by entering "sentence diagramming" into a search engine, or look it up in your old-fashioned high school handbook.

Be aware that some rules or conventions of punctuation or usage might vary somewhat among authorities or even among disciplines, and they can change over time; moreover, conventions used in American English sometimes differ from those of British English. Follow the requirements of your university or publishing journal. (And be aware that many common practices in journalism or business communications are inappropriate or unacceptable in formal scientific writing.)

Next to knowing and applying the fundamentals, you must exercise them regularly. Your writing—and your pursuit of a degree—will not progress well if you write only sporadically or if you rely on binge writing. You must develop a routine, and begin it early in your academic career. Do not wait until you are "ready"—or forced—to begin writing. Chapter 3 gives you plenty of things to write about, so get up in the morning and write.

The following guidelines address some of the most common difficulties college students have in scientific writing.

Paragraphs

- Begin each paragraph with a topic sentence, then develop that topic in the remainder of the paragraph. If you change topics, start a new paragraph. Try writing the topics in the page margins of your rough draft, or highlighting key words in your word processor. Remember to remove the highlighting from your final version. (Note that in nonscientific writing—fiction for example— the topic sentence may come anywhere in the paragraph.)

- Ensure that paragraphs really do exhibit development. If you make a general statement, back it up with specific examples or data. This is especially important in writing literature reviews.

- End each paragraph with a sentence that completes the thought and leads logically to the next paragraph. If the next topic is not logically related to the one just completed, one or the other could be out of sequence or a heading break might be needed. Alternatively, the difficulty could be that you really don't know what you're trying to say. Consider all possibilities.

Sentences

- Write complete sentences, using the standard structure of subject–verb–object.

- Avoid wasteful words. Make every word and sentence count (see Strunk and White, 2000).

  Wordy: "It has been shown that bovine serum albumin can be used as a suitable protein standard (Cowhorn, 2011)."

  Better: "Bovine serum albumin is a suitable protein standard (Cowhorn, 2011)."

  Weak: "There is evidence to show that ocean sediments are stratified."

  Better: "Evidence shows that ocean sediments are stratified."

  Best: "Ocean sediments are stratified (Shark, 2010)."

- Ensure that the word you *intend* to be the subject of the sentence really *is* the subject and not the direct object or some other element. *Do not begin a sentence with an author's name,* even though you commonly will see this construction in respected published works.

  Weak: "Cowhorn (2011) determined that bovine serum albumin is a suitable protein standard."

  Better: "Bovine serum albumin is a suitable protein standard (Cowhorn, 2011)."

  The first example sets you up to talk more about Cowhorn, which might be appropriate if you were writing a biography of Cowhorn. The second example sets you up to talk about bovine serum albumin, which presumably is the intended focus.

- Try to write more in the active voice than in the passive voice. In active voice, the intended subject does the action, rather than being acted upon (in which case, it's no longer the subject but the object). Do not compose a sentence in the active voice if that construction might divert attention from the intended focus.

  Passive voice: The questionnaire was responded to by 80 people

  Active voice: Eighty people responded to the questionnaire.

- Avoid writing long strings of prepositional phrases.

- Maintain parallel structures in sentences. For a sentence to be coherent, its parts must be connected and logically sequenced. The same rule applies to sequential sentences in a paragraph. Do not change the order of related points from one sentence to the next.

  Unacceptable: "An atomic absorption spectrophotometer was used to measure copper, and methane was analyzed by gas chromatography."

  Acceptable: "Copper was measured by atomic absorption spectrophotometry, and methane was analyzed by gas chromatography."

  Which is more important in the above examples, the substances analyzed or the methods of analysis? Whichever it is, place the more important factor first (i.e., make it the subject).

- Avoid false coordinations. Do not place unrelated clauses in the same sentence, and do not present a principal clause and a subordinate clause as if they carried equal weight. Place subordinate clauses or phrases in subordinate positions.

  Here are some examples from everyday language that might help emphasize the importance of precision in scientific communications.

  Unacceptable: "I went to Los Angeles, and I saw the Lakers play."

  As presented, these are either two separate, unrelated activities, or they are related activities that are equally important; either case is doubtful.
  Acceptable: "I went to Los Angeles, where I saw the Lakers play."

The second part of the sentence is subordinate to the first part. Going to Los Angeles was more important than seeing the Lakers play.

Acceptable: "I saw the Lakers play in Los Angeles."

In this case, seeing the Lakers play is the more important point; Los Angeles is incidental. (Note that this short, seemingly clear statement still is somewhat ambiguous even though it has no false coordination. Who really was in Los Angeles—you, the Lakers, or both? You could have been in Buffalo watching a Lakers home game on TV. How might you reword the sentence to ensure precision?)

Acceptable for emphasis, but usually not in scientific writing: "The Lakers won the championship in Los Angeles, and I was there!"

As these examples suggest, practicing precision in everyday language can be good practice for scientific writing. (Just don't get too hung up on it and take the fun out of everyday talk.)

- Complete all comparisons.

  Unacceptable: "Women need more iron in their diets."

  More than what? What is being compared?

  Acceptable: "Compared with men, women need more iron in their diets."

  Acceptable: "Women need more iron than zinc in their diets."

  Acceptable: "Women should increase the amount of iron in their diets."

- Remember that a modifier at the beginning of a sentence refers to the subject of the sentence.

  Unacceptable: "Using a tow truck, the car was pulled from the ditch."

  Did the car really use a truck to pull itself from the ditch? (Note that even though the verb "was pulled" might suggest that something or someone else operated the truck, this point doesn't override the rule. "Car" is still the subject and "Using a tow truck" is still its antecedent modifier, and the sentence is still ungrammatical.)

"Where were you that week in the eighth grade when Miss Write-right was teaching subjects, verbs, objects, and prepositions?"

Acceptable: "Using a tow truck, the emergency crew pulled the car from the ditch."

Acceptable: "A tow truck pulled the car from the ditch."

Punctuation

- *Colon.* Do not use a colon (:) (or any other punctuation mark) to separate a verb or a preposition from its object(s). Except in headings, the sentence preceding a colon should be grammatically complete.

  Unacceptable: "University students like to: eat anchovies, go to bed early, and spend Spring break writing papers."

  Acceptable: "University students like to eat anchovies, go to bed early, and spend Spring break writing papers."

  Unacceptable: "The study included: diameter, weight, color, and firmness."

Acceptable: "The study included diameter, weight, color, and firmness."

Acceptable: "The study included four parameters: diameter, weight, color, and firmness."

- *Semicolon.* Most often, the semicolon (;) is a mark whose function is intermediate between that of a comma and a period; think of it as a semiperiod. Use a semicolon in a compound sentence to separate closely related independent clauses that are not linked by a coordinating conjunction (as in the previous sentence). Here are three more common uses of semicolons:

  Place a semicolon before a conjunctive adverb (e.g., therefore, otherwise, consequently, nevertheless, accordingly) that links two independent clauses. Place a comma after the adverb.

  Insert a semicolon between sets of linked items in a complex series of data, conditions, or events when items in the series contain internal commas or other structural breaks (e.g., calcium, 36 mg kg^{-1}; magnesium, 4 mg kg^{-1}; potassium, 2 mg kg^{-1}).

  A semicolon *may* be placed before a coordinating conjunction (i.e., and, or, but) that links independent clauses if the clauses contain internal punctuation or if a strong separation is needed. Note that this usage varies among authorities and experienced authors. In straightforward cases, a comma coupled with the coordinating conjunction usually will suffice, but if the sentence could be misread, place a semicolon before the conjunction. Here are a couple of examples, the first with a comma and the second with a semicolon before the coordinating conjunction.

  "As expected, data for carbon, hydrogen, and aluminum fell within the normal range for each, but values for potassium, calcium, and magnesium were well below normal."

  "Despite the obstacles, we completed all physical, chemical, and microbial experiments and statistical analyses; and we met our reporting deadline, which ensured continued funding for the following year."

Word Choice

- Use of first person is acceptable. It's certainly better than the old practice of referring to one's self in the third person; nonetheless, place yourself in the background.

- Avoid use of "you," unless you really mean any person who might read your report; otherwise, identify the relevant person or group.

   Unacceptable: "You must ensure proper sequence within the algorithm."

   Acceptable: "Programmers must ensure proper sequence within the algorithm."

- Avoid beginning sentences with indefinite words (e.g., "There are...," "It is...").

- Avoid "look at" when you mean investigate, study, or research.

Variety

Often, students are told that they can avoid dull writing by varying their sentences. Too often, this advice results in disjointed paragraphs, incomplete sentences, wordiness, confused subjects, incompatible mixes of active and passive voice, strings of prepositional phrases, faulty parallelisms, false coordinations, improper comparisons, misplaced modifiers, misused punctuation, improper word choice, and general confusion.

Don't try to put variety in your writing simply for the sake of variety. Variety should have a purpose, it should be controlled, and it should not confuse the reader. Write first for clarity. Learn and practice the basics—variety will follow.

Free Writing

When your thoughts feel jumbled, your inspiration lies dead, and writing seems impossible: free write. When clarity suddenly strikes, an idea jumps into your head, and your enthusiasm bubbles: free write. For the times between: free write. Do it regularly. Make it part of your routine, and you will make progress.

In free writing, you focus your full attention and energy on moving thoughts quickly and smoothly from your brain to paper. The thoughts need not be organized or even coherent. You simply write them as they come—no sorting, no weighing, no refining. It's a bit like dumping clutter from a box—whatever is at the edge falls out first and the rest follows. Once the thoughts have spilled onto paper, you can sort through to find those that are worth keeping, some of which will be gems you never knew you had.

If done properly, free writing is intense. It requires uninterrupted, short-term concentration and proceeds at a steady pace—a literal stream of consciousness that summons your thoughts, frustrations, and inspirations to the fore. In some sense perhaps, these are forced from your right brain so quickly that your left brain never has a chance to analyze, evaluate, or filter them.

Here's how to do it:

1. Choose a quiet, preferably uncluttered place and a time when you will not be disturbed for about a half hour. Mornings seem to be best for most people. (Even if you consider yourself a "night person," reserve time first thing in the morning for free writing. Just get up and do it.)

2. Get out a few sheets of clean paper and a reliable pen. A notebook dedicated to these exercises serves well. (OK, use your word processor if you want. Just don't cheat the process.)

3. Set a time limit for the session—ten to twenty minutes is plenty.

4. Sit and stare at the blank paper for about a minute while you **reflect on your topic.**

5. Begin writing. If you can't decide how to begin, just write whatever is in your mind at the end of the minute. Once you start, write steadily for your allotted time, never diverting your attention from the paper. Do not hesitate, do not look up, do not reread, do not edit, do not cross out, do not struggle over word choice or spelling or punctuation. Do not inhibit digressions; let them flow. Just write rapidly forward, putting in squiggles if you can't think of a word.

Stop when your time limit is up. Rest for a few minutes and think about what you wrote. What old thoughts and what new ideas came out? What connections and patterns emerged? Did you produce any particularly good phrases or sentences? Consider the digressions you took. What are they? What do they mean to you and to your topic? Should they be developed? If so, dedicate another session to them.

You will see lots of rubble in your writing, but you'll also find nuggets among the gravel. Sluice them out and polish them. Before you began, the rubble was in your brain, now it's on paper. Before you began, your left brain struggled to identify, arrange, and plan the future of the useful ideas, but it had difficulty picking them out. Before you began, inhibitions might have concealed and suppressed some of the more deep-seated or even secret thoughts; now they're out for honest recognition and appraisal.

Free writing brings out your creativity, but it also can help pull you through the downers. If you free write regularly four times per week, you can find yourself making progress even when you're in a slump. Bear in mind that no one else needs to see what you write. It's something you do for you.

Preparing a Literature Review

What Is a Literature Review?

A literature review is a formal written account that summarizes and integrates relevant information found in various books and articles. It's different from an annotated bibliography, which is a collection of individual references with each reference summarized and critically evaluated. Whereas each reference in an annotated bibliography stands alone, selected contents of references in a literature review are melded into a coherent narrative. Literature reviews can take several forms, depending on their intended use. Here are five types:

1. **Text books.** These range from introductory to advanced but still are the most general. You know about them.

2. **Peer-reviewed stand-alone articles found in professional publications.** These review papers summarize research findings on a topic of interest to the readership. Authors assume that readers know the fundamentals; hence, articles tend to be more detailed and scientifically sophisticated than introductory textbooks or popular review articles. They are judged for currency, accuracy, and comprehensiveness by a panel of experts before publication.

3. **Review articles written for popular scientific publications.** Compared to the second type, these are directed toward a broader audience, one having scientific interest but less sophisticated knowledge in the field. Compared to professional reviews, these are technically less advanced and demanding but still highly informative. They commonly are written at about the level of an introductory text book.

4. **Short bits of background information and specific evidence that are woven into the introduction and other parts of original research reports.** These are the most common type of literature review. In the old days, prior to about the 1960s, scientific papers included literature reviews under separate headings. The integrated format was introduced to save space in the journal.

5. **Chapter in a graduate thesis.** In a traditional thesis or dissertation, the literature review takes two forms. First, the formal literature review is a chapter unto itself. Its form and purpose are similar to a peer-reviewed stand-alone article (no. 2 above). Second, additional background information, technical references, and specific evidence are woven into other chapters (similar to no. 4 above), including the introduction, methods, and discussion, but normally not into the results chapter.

Here's what the literature review chapter in your thesis should accomplish:

- accurately and succinctly present the main points or ideas of relevant published works
- support general statements by direct evidence and specific examples, including data
- properly cite all references

Your job in writing a literature review is much like that of a textbook author, who must seek out, read, understand, and summarize numerous scientific articles and then present the most important information, including pertinent facts and details, in a coherent narrative. It is like writing a term paper on the subject of interest and incorporating it into your thesis.

Often you must go beyond the immediate scope of your work to learn important general information needed to get the project going and to reveal specific clues that later will help you interpret your results. For example, if your study is on the flight of bumble bees, you might need to explain the mechanics of flight and the anatomy of bees. Obviously, you must look up information on bees, but don't limit your efforts to bumble bees: look up honey bees and wasps, too; and while you're at it, check out humming birds, hawks, and helicopters.

Although the literature review summarizes and integrates what people have learned about the subject, it may contain critical comments when they enhance the thesis. Any criticism must focus on the work, not the workers, and it must never be based on unsupported personal opinion or innuendo. If used effectively, as tools to prompt and guide your progress, the initial criticisms should be either strengthened and discussed in your thesis or discarded.

Why Write a Literature Review, and What Should It Accomplish?

Here are some of the main reasons for preparing a literature review:

- The review helps you to find out if your problem, or some part of it, has been solved.

- It will help you to learn pertinent laws, theories, facts, and interpretations that can provide a foundation for your work. *Reading familiarizes you with the subject; writing about it ensures that you understand it clearly.*

- It summarizes the state of knowledge on the subject for your reader (and for you!).

- Gathering and organizing background information on the subject allows you to identify gaps and inconsistencies in knowledge and to form new questions. Each of these questions then can be assigned a priority.

- Preparing the review will help you define your problem or goal, recognize and select key factor and variables, and specify the objectives of your work.

- The review provides information needed to develop methods, establish a suitable design, select proper statistical and analytical tools, collect the proper data, and identify specialized equipment.

- It makes you familiar with the names and interests of experienced researchers and problem solvers and leads to additional sources of information.

- It shows you techniques of presentation.

- It helps facilitate interpretation of your results. Documented examples and data from past work can serve as reference points or measures for comparing and interpreting your own work. Don't neglect this opportunity to make the literature review work for you.

- Writing the review teaches you skills needed to deal with future questions.

- It shows your reviewer that you've done your homework. This is especially important in securing funding or satisfying a client (including a graduate committee).

- Writing a literature review requires that you bring together and evaluate ideas and facts from diverse sources, rather than considering papers in isolation from one another. Doing this with commitment forces you to methodically examine the logical consistency within and among interpretations by various scientists and authorities—something every graduate student should learn to do.

- It teaches you how to teach yourself, which is another big reason for being a graduate student!

The reasons for writing a literature review also suggest several things that the review should accomplish. Use the above reasons as a checklist. Does writing your literature review help you to accomplish each of the points mentioned? If so, you're doing a good job!

When Should the Literature Review Be Written, and How Long Should It Be?

Begin reading for the literature review as soon as you ask, "Where do I start?" Then write. Update your work often to keep abreast of new developments and edit out material that becomes irrelevant as you change direction. Continue revising until you complete the final thesis.

Although the length of a literature review cannot be predetermined, it might run 10–30 pages. The only way to determine the proper length is to write it. A brief review does not show the student's depth of understanding (or perhaps it does) and normally does not provide adequate information to allow full and proper interpretation of results. An unnecessarily long review suffers from inadequate editing.

What Topics Should Be Included?

The example mentioned above about the flight of bumble bees suggests some of the logic to use in selecting topics. Keep in mind the purpose of your work, the complexity of the subject, the knowledge level of your audience, and your own knowledge of the subject. Begin with an exploratory search for background information, and develop a list of key words to guide you to more specific topics.

Also, remember that you may need to search beyond the scope of your project and extrapolate back to it. The chances are slight that you will find a report that covers your subject exactly—if you did, you wouldn't have a research topic.

What Are the Various Types of References, and Where Are They Found?

References can be primary or secondary. A *primary* reference is an original report of original research. It contains findings that are generated by the author(s) specifically for the paper; nonetheless, it also may contain secondary information in its literature review. Primary references include original reports in scientific journals, experiment station publications, some conference proceedings, and other comparable sources. Normally, the majority of references in a thesis should be primary.

Be aware that even the most recent primary publications might not contain the most up-to-date information because of lag time for reviewing, editing, and publishing. The lag time can be shortened in online journals; nonetheless, proper reviewing and editing remain time consuming. If the most recent information is critical to your project, you might contact authors directly to find out what they've done since their last publication, but read their papers first.

Secondary references contain data and other information that usually were not generated by the author(s) and that previously appeared in a primary article. Secondary references include textbooks, review articles, most monographs, and most abstracts published by abstracting services.

Although secondary information is acceptable in a thesis, you always should attempt to consult and cite original papers. Not to do so is to trust that the second author has presented the facts accurately, maintained the proper context, and provided all the details you will need. In addition, citing the reference properly often is more cumbersome for secondary than for primary sources.

Here are some steps for beginning your literature search:

- Do your homework. Read text books and other secondary sources to gain background knowledge and direction.

- Compile a list of key words for more detailed study; include your anticipated key factors and variables.

- Consult online catalogs, databases, reference services, science-specific search engines, and related electronic sources. Many are available, nationally and internationally. Some are accessible by subscription only. Check with a reference librarian at your university and with experienced students for the most helpful websites. Sites and URLs change periodically, so they're not listed here.

- Consult with your advisor and others for leads.

- Get to know the journals in your field. Pick up a paper copy and thumb through it as well as scanning the electronic version.

- When you find a usable article, check out its references and more recent papers that have cited it.

- Go to the library—physically. Get to know it and get to know the librarians. Despite the convenience of the internet, this is still an excellent place to look for information. Check out the reference and current periodicals rooms, government documents, the abstracting services, interlibrary loan, and of course the stacks. Plan on spending an hour or so learning to use the resources— then use them. Your time will be well spent.

Be aware that although electronic databases are helpful, they also are incomplete; you will not be able to do your entire literature search by computer. Neophytes often consider the computer search to be their first and last line of inquiry, thinking that it will do all the work for them. Experienced people know that although a well-planned computer search can yield respectable returns, it will not reveal the nuggets that can turn up by the old-fashioned dig-it-out-yourself method.

How Does One Organize and Prepare a Literature Review?

The organization of a literature review depends largely on its purpose or type (see the five types of literature review described earlier). As a chapter in your research plan (and later in your thesis), the literature review should pick up where the introduction leaves off. It should probe deeply into your chosen subgoal.

The literature review probably will be the most difficult writing in your thesis. You must synthesize a coherent narrative from disjointed and often difficult information gleaned from diverse sources that exhibit variable and sometimes cryptic writing styles. Most novices are intimidated by the job, and few experienced researchers do it really well. The good news is that once you learn to write a well-crafted literature review, other scientific writing will be comparatively easy, and the skills you learn will serve you well throughout your career. The following suggestions combined with the earlier reading and writing guidelines in this chapter will help relieve the intimidation and ensure a good-quality product.

First, set reasonable immediate goals, and set them frequently. A reasonable goal is to find and make notes on one or two references, or to write 200 to 500 words. A reasonable goal is not "to work on my literature review" or to set an activity requiring several days.

Begin compiling lists of potential topics, key words, journals, and authors. Use these lists to help guide the rest of your search and direct your writing. The key words will suggest more topics, and many key words will require definitions, explanations, and interpretations. Don't spend too much time on this activity; as soon as you have a few key words, commence reading and writing.

Although you can organize the literature review in various ways, you might begin by presenting detailed and documented background information on your research problem; then organize around your potential key factors and variables. (You won't be able to finalize these until you've progressed with the literature review; so it's a back-and-forth process.) Organize by topic, and use topic headings—do not organize according to authors or the sequence of papers you read. The point is to *be informative and make the content flow.* If a single article contains different types of useful information, separate the types, then group each type with related details from other papers. (Be sure to keep track of all the sources so that you can cite and reference them properly.) Segregating, strengthening, and reorganizing the diverse points at this time—a bottom-up approach to building your literature review—will greatly ease the task of integrating the information into a coherent narrative. This approach also will help ensure a strong framework and an informative, defensible product that can be readily edited and amended.

In keeping with the bottom-up approach, make concise notes about the scientific papers as you read them. Don't print or photocopy each paper and then try to write your literature review from the copies. This practice is cumbersome and inefficient; it increases the difficulty of your job. Instead, use note cards (the 4-in. by 6-in. size serves nicely) or compose card-sized minidocuments in your computer. Each card or minidocument should contain only relevant, closely related points, including representative data and specific examples whenever possible. Using this technique, you might generate several cards or minidocuments from each article. (For working directly on your computer, a large monitor, or dual monitors, is almost indispensible for arranging multiple minidocuments in a single view.)

To improve efficiency, you might want to use annotation software that allows you to insert comments and transfer selected points from online articles to the software or to your usual word processor. But as helpful as this type of software might seem, it cannot properly collate diverse bits of information to fit your particular purpose, nor can it express the information in your words. (Remember, you are preparing a literature review—an annotated bibliography is merely a step along the way.)

For these reasons, you should learn the traditional note-card technique, which has been used successfully by generations of scientists. It works exceptionally well for aggregating related details from diverse sources; moreover, it can help you uncover obscured relationships and meanings, generate new ideas, and exercise your skills in organizing thoughts as well as facts. Once you've learned the process, feel free to try other methods, but don't discard what you learn by using note cards.

Here's what to include on the cards:

- Full and proper reference of the paper *precisely* as it must appear in the final copy of your thesis or professional paper (After the first card in a set, full references are not necessary—author-year will do. If you also are using referencing software, you need only the author-year citation at the top of each card.)

- Number of references cited in the paper (This information is optional, but it can be helpful later.)

- Call number and location of the publication in the library, or internet URL (Again, you might want this later.)

- Salient points in your own words (Write in outline form. Avoid direct quotations unless you surely must use them. In using quotations, be certain to enclose them between quotation marks and to note the page number from which the quotation was taken.)

- Direct evidence in the form of specific examples and data that support the salient points (Go beyond mere generalities. You might even want to photocopy a useful table or graph and tape it to the card, or if you're reading an electronic version, download it to your computer. Either way, you probably will present only a summarized version in your paper.)

- Card number in the upper right corner if more than one card is required (e.g., 1-3, 2-3, 3-3 for three cards in a sequence).

Once you've compiled notes from several references, lay out the cards on a table and arrange them in a logical sequence. Fill in topic headings and transitions on additional cards. If one card has multiple bits of information that belong in different parts of your paper, make multiple cards. After the cards are laid out, ask a colleague to critique your organization and content. Finally, stack the cards in proper sequence and begin writing your literature review from them.

If you become perplexed and bogged down in the writing, you probably don't know what you're trying to write. In other words, if you find yourself saying "I know what I want to say, I just don't know how to say it," you probably really *don't* know what you want to say. (People writing in a second language could be an exception.) Back up and reevaluate the immediate topic and your knowledge of it. Reread the papers you're trying to cite, and be sure you understand them. Check the definitions of all uncertain words. Most important, don't short-circuit the spirit of the note-card technique, even if you do not follow it to the letter.

What about the Writing Style?

Keep it simple and straight forward. Heed the "Hints for Scientific Writing" offered earlier in this chapter. Follow the standard rules found in nearly any good book on English grammar, composition, and style (Strunk and White's *Elements of Style* serves nicely).

Separate major subjects by headings. The headings will help you organize your work and set reasonable goals, and they will help your reader.

Begin each paragraph with a topic sentence, then develop that topic in the body of the paragraph. In writing the topic sentence, pay particular attention to your choice of subject and verb. The subject of the first sentence should set the topic of the paragraph. Note that many scientific authors do not follow this rule; the result is that the reader becomes distracted from the main point, and the reading becomes more difficult.

Here's a sentence that has a poorly chosen subject: "Geppetto and Pinocchio (2011) found that wood fibers soaked in chicken fat for 180 days became soggy." What is the subject of the sentence? What do you think *should* be the subject? Can you name at least two alternate subjects that the writer might have chosen? Rewrite the sentence in two different ways, each having a different subject. Under what condition would the present subject be appropriate? Complete this little exercise before reading further. (You might recall that a similar example was presented earlier under "Hints for Scientific Writing.")

OK, now that you've answered the questions and rewritten the sentence, read on. The subject is Geppetto and Pinocchio, but the writer's main interest probably concerns wood fibers or perhaps chicken fat. Either substance could have been made the subject; instead each was relegated to be the object of a preposition. Here are a couple of ways the sentence could be improved:

Emphasis on wood fibers: "Wood fibers became soggy when soaked in chicken fat for 180 days (Geppetto and Pinocchio, 2011)."

Emphasis on chicken fat: "Chicken fat renders wood fibers soggy after 180 days (Geppetto and Pinocchio, 2011)."

The original sentence, in which Geppetto and Pinocchio is the subject, would be appropriate in a biography of the two scientists, or perhaps in contrasting their work to that of competing scientists.

The above exercise demonstrates that a writer should place emphasis where it belongs. In writing a literature review, that normally means focusing on what was discovered, rather than on who made the discovery. Note also that with focus placed on the actual intended subject, the structure is set up for a smooth transition to the next sentence.

In addition to understanding the difference between the subject of a sentence and the object of a preposition, you also should distinguish the subject from the direct object. Normally, the subject carries out the action and the direct object is acted upon. Keeping this point in mind will help you to write in the active voice.

Establishing Credibility

Every writer, especially those who write nonfiction, must establish credibility with the readers. Readers must feel confident that information is presented honestly and accurately. Journalists strive for credibility by supporting their articles with direct quotations from known or supposed authorities. (Unfortunately, we never know how much the quoted authorities really know.) Although this technique is widely used outside of science, it does not work well in sophisticated scientific writing. It looks like a crutch. Don't use direct quotations to present evidence in your research plan or thesis.

Quotations in scientific papers and theses have two negative consequences: (i) they divert focus from the message to the messenger, and (ii) they give the impression that the writer lacks the confidence or skill to put the information into his or her own words. Excessive use of quotations goes further: it suggests incompetence.

In scientific writing, credibility and authority are established by citing specific, direct evidence from peer-reviewed sources and by including complete, correct references. Specific examples, data, and other forms of direct evidence lend credence to your testimony. They show that you have done your homework. The reader knows that you actually read, or at least looked at, original research reports. Direct evidence also shows that your line of reasoning goes beyond a mere rationalization.

Should you *never* use direct quotations? Not quite, but use them rarely—solely for a specific purpose that only a quotation could achieve.

What Is the Proper Format for Citations and References?

Follow the guidelines given under "Citations and References," in Chapter 3. Use the author-year format for citations in the body of your paper (e.g., Geppetto and Pinocchio, 2011). You might have to change these to the number format for publication in a scientific journal, but for now the author-year format will be more convenient.

Literature review references should be presented together with other references (e.g., from your methods and discussion sections) at the end of the research plan. Follow the standard format used by professional journals in your field or required by your university program. For most fields the format used in this book will be acceptable or at least close.

Check into a computer program that automatically formats and revises references for you. If you're unsure about these, ask experienced graduate students; one of them will know what's latest and best in the market.

Avoiding Plagiarism

In writing, plagiarism is a form of fraud or intellectual theft in which the writer presents the work of another person as if it is his or her own. As a violation of intellectual and academic integrity, plagiarism can have serious long-term consequences, including loss of reputation and funding, university-imposed punishments, and sometimes even criminal charges.

Plagiarized work can include data, a sentence or phrase, or even an idea—any type of intellectual or creative property that can be traced to an uncredited individual or group—regardless of how it is acquired or whether it is copyrighted or not. Plagiarized material can come from either print or electronic sources including computer files and internet postings. (Verbal sources also can be plagiarized, but these cases usually are uncertain and difficult to prove.)

Certainly you can present someone else's ideas or findings in your report, but you must give due credit through proper citations and references. Also you may use another author's wording provided the phrase or sentence is placed between quotation marks. If you borrow an idea from another source and put it in your own words, you still have plagiarized if you don't give the original author proper credit for the idea. The same rules apply to your own original ideas and wording if your work was previously published and copyrighted.

Plagiarism normally does not include widely reported or commonly known definitions, facts, and events; provided they are expressed in your own words. Nonetheless in most cases these should be cited and referenced at least for documentation purposes. Numerous internet sites, including those maintained by universities, provide additional details and examples of plagiarism and how to avoid the problem. Check your university catalog for the plagiarism policy at your institution.

Writing for Scientific Publication

Your research is not complete until it is published where other people can read about it. If you've done a good job of writing your research plan and keeping it updated through the final stages of your work, writing the publishable paper should not be difficult. The sooner you get started, the sooner you'll see your name up in lights!

The Process

The overall process for publishing in a peer-reviewed scientific journal (also called a refereed journal) typically runs something like this:

1. The authors assess their work, do some background inquiry, and make decisions about what, where, and how to publish.

2. The authors write the manuscript, and the corresponding author submits it to the journal editor. With hard-copy submissions, three additional copies for reviewers usually are required.

3. The editor checks the manuscript's format and content, then sends copies to expert reviewers. Following evaluation guidelines similar to those presented earlier in this chapter, reviewers comment on the paper's strengths and weaknesses and recommend whether to publish or not.

    Depending on journal policy, the reviews might be open, single-blind, or double-blind. If the review is open, the names of authors and reviewers are known to all; if it's single-blind, the authors' names are withheld from the reviewers; and if it's double-blind, only the editor knows the names on each side.

    In addition to comments, reviewers typically make one of four recommendations: (i) The paper is in great shape, publish it as is (a rare recommendation). (ii) Publish without additional peer review after authors make corrections or minor improvements. (iii) Return the paper to authors for significant revision, and allow it to be resubmitted. Require full peer review of the revised manuscript. (iv) Release the manuscript to the authors (a polite way of saying reject it). Manuscripts might be rejected for any of several reasons, including redundancy with papers already published, fatal flaws in methodology, weak or inadequate data, or inappropriate content for the journal.

4. The editor informs the corresponding author about the reviewers' comments and recommendations and provides instructions or suggestions for the next step.

5. If necessary, the authors revise the manuscript and resubmit.

6. When accepted, the manuscript enters the queue for next available publication.

Get Ready

Publishing a scientific paper requires preparation and a few decisions. Discuss each of these with your major professor:

- Decide whether your study can be split into more than one publication.

- Identify the most suitable journal(s)—the one(s) that most probably will publish your work and reach your audience.

- Decide on authorship. The greatest prestige and workload normally go to first author. The first author usually is the one who corresponds with the journal editor.

- Determine whether the publication will be online or on printed paper or both. Find out about special requirements for electronic or paper submissions and publications and about acceptable software for electronic submissions.

- If you hope to publish through a professional society, look up the membership requirements and privileges.

- Check on publication costs—that's right, in most cases you pay them.

- Find out about the review process and typical review time.

- Check on copyright ownership and how reprints are made available.

- Read the journal's instructions to authors. You can find them on the journal's website or in a printed version. Note the formatting requirements for citations, references, tables, figures, and overall layout. Tables and illustrations often are placed together at the end of the author's manuscript. Be sure to use proper units of measure. (The physical sciences usually adhere to the SI system, *Le Système International d'Unités*). Follow all instructions precisely.

Write the Paper

Different people approach the writing of a professional article in different ways. No single approach is best for everyone, but still you probably could use some suggestions to get started and gain momentum. (For more in-depth coverage of publishing in scientific journals see Coghill and Garson, 2006; Valiela, 2009; and Gladon et al., 2011.)

For most people in their early careers, the greatest challenges are in reducing several years of work to a few pages of journal manuscript and getting it done. Some writers find the job of paring the unessential parts to be particularly trying—especially when those parts were painstakingly produced. But once you've trekked this far along the research path, you shouldn't allow extra baggage to drag you down. And you shouldn't try to hand it off to your reviewers or readers. Most of them will have hiked a similar trail—reviewers won't be impressed, and readers probably wouldn't read it anyway. Later, if you decide it's worth recycling, you can go back and pick it up.

Break the work into sections, as you did for the research plan (Chapter 3). Don't try to write any section completely before beginning the next section; don't even attempt full paragraphs at the start. If you do, you likely will find yourself losing thoughts and direction and ending up in an editing quagmire. Instead, begin with single statements or small units of outlined information, and separate them by headings. Give the following steps a try:

1. Write a single statement that describes the problem of your research. Make it succinct but informative. Don't simply write what your work is *about*; write what it *is*. Assume that your audience might not be aware of the problem but that they readily can comprehend it.

2. State the importance of the research. To whom is it relevant and why?

3. Write your objectives in a single sentence each or in outline form.

4. Without elaborating or incorporating extensive data (that will come later), write a statement of each significant result. Make each statement informative, definite, and brief. Do not include interpretations. Refer back to "Results vs. Interpretations" (Chapter 2) and to "Anticipated Results" (Chapter 3), as well as to your research plan.

5. Write each of your major conclusions or interpretations in a single sentence each. Do not write anything that does not follow from your results or that does not pertain to the points of relevance in step 2. If your writing does begin to deviate from those points, one or the other is faulty and should be adjusted.

6. Make up a title of no more than a dozen words. Don't struggle with it or lock in on it yet. You probably will revise and shorten it several times.

7. Outline one or two ideas for future studies that your work has revealed.

8. Back up your computer files after every session.

Steps 1–7 construct the skeleton of your paper. You still need to add the methods, abstract, and references, each of which will come shortly. So now let's proceed with the in-depth writing.

Results: Back up to step 4 and begin writing detailed results. Construct tables and figures of data to be included in the manuscript, focusing on significant points but not omitting essential supporting documentation. Continue the narrative from the statements you wrote in step 4 and in your research plan. Write only about your work, without mentioning other studies or what your results might mean. Again, refer back to "Results vs. Interpretations" (Chapter 2) and to "Anticipated Results" (Chapter 3), as well as to "Hints for Scientific Writing" in this chapter.

Methods: Next write the methods section. This should be a fairly straightforward task, which most authors find less demanding than other sections (Gladon et al., 2011). Methods are good to work on when you get stuck somewhere else. Unless they are relevant to your findings, leave out all the zigs and zags and misstarts that you experienced. Be sure to include proper citations.

Introduction: After the methods, write the introduction from your statements in steps 1, 2, and 3 above. Pose particular questions that your study addresses. In addition, note that modern journal articles usually weave a brief literature review into the introduction, rather than presenting it in its own section.

Discussion: When you finish the introduction, you should be reasonably well prepared to write the discussion section, which differs from results in that it analyzes findings and plays up their significance. The discussion addresses "so what?" and "who cares?" types of questions. It includes interpretations and conclusions along with suggestions for future investigations. It should *not* introduce new material or tangential ideas that do not directly pertain to or follow from points presented earlier in the paper.

The discussion probably will be the most challenging part of your paper to write, but you can compose it solidly and artfully if you construct it around a few core points from the results. Once you've chosen those points, connect them to the problem and potential applications presented in the introduction. Answer any questions raised in the introduction, and discuss whether or not your objectives were met and hypotheses rejected. Show how your results relate to the findings of other investigators as outlined in your literature review, again using proper citations. If your study brought to light any possibilities for future work, say so.

References: Many writers construct the references (sometimes called literature cited) section intermittently as they compose other parts of the paper. However you do it, follow the journal's formatting instructions precisely. Be sure that any nonoriginal work—including your own if it was previously published—is properly cited (See the section "Avoiding plagiarism." earlier in this chapter.). Check that each citation is properly referenced and that each reference is properly cited.

Abstract and Title: Last, write the abstract and finalize the title (refer back to "Abstract," in Chapter 3). Word choices here are particularly important because these words will be used by indexing services and picked up online by word finders; moreover, the title and abstract often are duplicated and published separately from the whole paper.

Arrange and format everything properly for the journal, and that's it—almost. Your wrap-up steps are to polish the writing, proofread the manuscript, and give the whole thing your own critical review. Don't forget to do a final spell-check and to ensure complete consistency between citations and references. Reviewers detest these types of oversights.

Now you may submit the manuscript, pay any page charges, and await the reviewers' comments and the editor's decision. And when your manuscript finally goes through—congratulations!

CHAPTER 5

Keeping Track of Things

What's This Chapter About?

Notebooks and diaries

What's in This Chapter?

The Research Notebook

>   Mechanical Guidelines
>   Content

Your Personal Diary

Free-Writing Notebook

I think I'm losing my mind.

Comment from a graduate student who shall go unnamed

The Research Notebook

Keeping a proper, up-to-date notebook of all findings and activities is an integral part of any valid research effort. The notebook, combined with the research plan, will be the basis of your thesis or technical report. The notebook should be similar to those required by many chemistry, physics, and engineering courses and by employers, especially those in competitive private industry or research and development organizations (Kanare, 1985). You should record in your notebook all physical activities, thoughts, data, calculations, and progress pertaining to your work. Diligent note keeping will increase your understanding of your work, decrease your time spent in report writing, and increase the quality of your reports. The notebook will become increasingly valuable to you as your work progresses.

You should have your notebook checked periodically by your thesis advisor and a fellow graduate student. Each reviewer should indicate the pages reviewed, then sign and date the last page. The signature and date signify that the person has read and understood the pages noted and that all entries were made prior to the review.

Why go to all this trouble? Isn't it overkill? Perhaps in your field and for your project it is. Some of you won't have to get this elaborate or protective. But if you're working in a sensitive area or on a high-profile, controversial project that could stimulate antagonistic responses, you'll need to be careful. Also, high quality control or ethical protection standards could require scrupulous record keeping. Scrupulous or not, you should maintain a research notebook.

All entries should be written directly and permanently into the notebook *at the time the work is done*, not later to improve neatness. All reasonable reviewers and courts of law understand that notebooks cannot be perfectly neat and orderly; nonetheless, notebooks should not be sloppy or illegible.

Adhering to the following standards and guidelines will help ensure that your work is efficient and properly defensible (Kanare, 1985). (Note: keeping notes in your computer is not enough; major professors, laboratories, and funding institutions typically require original hard-copy notebooks as described below. And often these are not yours to keep—they belong to the project.)

Mechanical Guidelines

- A notebook that is bound (glued and sewed), hard cover, 8 1/2 by 11 in., with cross sectioned pages (horizontal and vertical lines) works well for most scientific projects.

- Write in black ink with ballpoint pen. Make corrections by drawing a neat single line through material to be deleted; do not try to obliterate incorrect entries.

- Include the following information on the cover: your name, school or company address and phone number, your advisor's name and university address and phone number. Repeat this information on page one.

- Reserve pages 2–5 for the table of contents.

- Number pages consecutively in the upper right-hand corner.

- Properly date and initial each day's entry.

- Writing must be legible and grammatically correct, although outline form is permissible when appropriate. Active voice and first person are preferred. This will pay high dividends when writing the thesis.

- Entries must be clear, concise, and complete. Another person must be able to follow the logic and reproduce the process.

- Include proper units for all numerical entries.

- Materials may be attached to the page by good-quality transparent tape.

- Begin each major activity on a new page, and use appropriate descriptive headings for each activity.

- Number the pages and record activities front to back in the book as you progress. Do not skip pages to be filled in later; instead write "continued on page __", and on that page write "continued from page __." This technique helps ensure that you will not be accused of improperly altering data or recording information after the time you designate.

- Photocopy new material each week, and do not store the photocopies with the notebook.

Content

Include the following items for each major entry:

- Date, type of activity (e.g., thought, observation, field or laboratory activity), and its objective.

- Methods. Make them general and give the complete and proper reference if you are following published standard procedures. Give details of any deviations from the published procedures and full details of any unpublished procedures you use. Remember that another person should be able to duplicate your work from your notebook instructions. Sometimes drawings, diagrams, or flow charts are helpful.

- Data and observations. Include any appropriate tables and graphs.

- Calculation set-ups and as many examples of calculations as needed to make the manipulations clear.

- Set-ups for any statistical analyses of the data.

- Interpretations of results.

- Comments and personal opinions. These should be indicated as such (e.g., Comment: . . .; Opinion: . . .).

- References using proper format.

Your Personal Diary

No one will require you to keep a personal diary, but you will find the activity helpful. Use it to record your thoughts, opinions, emotions, moods, and outlook—some of which you would not want to put in a research notebook. A diary will help you follow your personal progress through the ups and downs, and perhaps most important, it will help you release frustrations and understand your feelings during times of stress. If you have difficulty recording your feelings, try writing letters to yourself—then keep them for yourself.

Free-Writing Notebook

If you free write regularly—say for 15 minutes, two to four times per week—you should keep a free-writing notebook. As explained in Chapter 4, it's a nearly painless way of pulling out buried thoughts and jotting down ideas before you lose them. Keep your notebook handy for capturing random inspirations as well as for regular sessions. You quickly will accumulate a good stock of revelations and mundane necessities as well as key phrases and even full paragraphs to use in your research plan and thesis. That's progress. Soon, you'll be able to look back on a trail of thinking patterns and progression. You'll find nuggets that you forgot about, and you can go back to old ideas without having to reinvent them.

Unlike your research notebook, a free writing notebook has no special rules. Simply get something that's easy to carry around and large enough for good service; a standard 8½-in. by 11-in. spiral-bound will do. If you work with lots of diagrams, you might want cross-sectioned lines. Even if you free write on a computer, you'll need something for times when you're away from it. Either way, date the entries.

CHAPTER 6

Presenting Your Work and Yourself

What's This Chapter About?

Presentations at professional meetings

Résumés, curricula vitae, and cover letters

What's in This Chapter?

Oral Presentations

    Preparing the Talk
        Title
        Opening
        Objective
        Content
        Transitions
        Visual Aids
        Closing

    Preparing Yourself for the Talk
        Physical Manner
        Voice
        Language and Speech
        Miscellaneous Suggestions

    Delivering the Talk
    Self Analysis

Poster Presentations

Résumés or Curricula Vitae and Cover Letters

    The Résumé or Curriculum Vitae
        At the Top
        Education
        Experience
        Honors, Awards, Licenses, and Certificates
        Languages
        Professional Organizations
        Publications and Presentations
        Professional Goals and Interests
        Hobbies and Other Personal Interests

    Cover Letters

Is this the party to whom I am speaking?

Lily Tomlin

Oral Presentations

As a graduate student in a well-regarded program, you will be required to give several oral presentations on various topics. Periodically, you will update your graduate committee on your research and overall degree progress. In addition, you will present seminars on your work and related themes to the broader faculty and other students. Near the end of your program, you will formally present and defend your thesis work. Later, as well as during your studies, you probably will speak at professional conferences, board meetings, or civic functions. In addition, you might be called upon to give expert testimony in a court proceedings, and if you go into teaching, you will speak to groups regularly.

Design your message and approach to fit the occasion, most important, the audience, the format, and the venue. Weeks before the meeting, inquire about audience interests, knowledge, experience, motivations, and attitude, as well as the anticipated attendance level. Find out the format, including time limitations, question-and-answer sessions, and degree of formality. Ask about the size and layout of the room, availability of audiovisual equipment, and any special requirements or limitations (e.g., some venues might accommodate only specific computer operating systems or programs).

Be aware that the speaker's intent and the audience's perception are not always congruent. As the speaker, your responsibility is to know your subject, but also it is to anticipate the possibility of miscommunication. At large professional meetings, listeners usually can't interrupt with questions or ask you to repeat a point. Doing your homework and preparing well are the most effective means of ensuring that the words you speak and the illustrations you show are the same ones the audience hears and sees.

Although many of the following suggestions pertain to nearly any type of oral presentation and audience, they are intended particularly for scientific presentations to scientific audiences—talks that primarily are informative rather than motivational, persuasive, or entertaining (Winn, 1997).

These speaking engagements might range from five minutes to an hour, with formal talks at professional conferences typically running 15 to 20 minutes.

While inexperienced speakers usually accept that a long talk calls for lengthy preparation, they mistakenly assume that a short talk requires proportionally less preparation. To the contrary, an effective, professionally delivered 15-minute presentation requires many hours to plan, sort out, construct, and rehearse. Often two or three years of diligent research that produces thousands of data points and scores of written pages must be condensed to a half-dozen simplified charts and a few dozen lines. After all your hard work, you might feel that 15 minutes isn't much. It isn't, but a professional talk is your big chance—prepare for it and you'll do well.

Preparing the Talk

Start early—several weeks in advance. The mechanical processes will take longer than you probably think, and you'll need time to cogitate and edit along the way. Begin with a framework. The delivery portion of a 15-minute talk should run about 12 to 13 minutes, with the time divided into three main parts: introduction (10%), body (80%), conclusion (10%). This will leave a little time at the end for questions and comments.

Here's a fairly standard format you might want to follow:

Introduction
- Title, author(s), and perhaps acknowledgments.
- Overall problem or goal and justification for the work.
- Objectives.

Body
- Highlights of relevant previous work by you or others. Alternatively, references to other work could be woven into other parts of the talk wherever appropriate.
- Methods outlined.
- Results.
- Interpretations and implications. Although interpretations and implications should be developed separately from results (see "Results vs. Interpretations," in Chapter 2), they may be presented in parallel with results provided the distinctions remain clear.

Conclusions
- This is your wrap-up, the message with which you want the audience to leave. It could be highlights of your major findings and their implications, including ideas for further research.

Prepare your talk from the inside out, starting with results, interpretations, and significance of your work. Begin creating computerized visual aids that will render these major elements effectively. Don't try to make anything perfect at this time; just put down the main points. You will edit and reedit several times before delivering the final presentation.

Next, work backward to the introduction: title, goal, and objectives. Ensure that the introductory elements are consistent and compatible with each other and with the results and interpretations. Move back and forth between and among elements until you get them right.

Keep in mind that some in your audience will be experts and devotees; they will understand and evaluate your every word and data point. Most, though, will attend with only peripheral interest; they will want to know the highlights and significance, but not much more—at least for now. Part of your challenge is to satisfy their immediate curiosity and hook their further interest. Also keep in mind that your goal is to inform a group of professionals about your contribution to their field—it's not to impress them with the huge amount of work you've done. Keep your talk straightforward (but not simplistic), informative, and relevant.

Before long you will realize that you have far too much to present in 15 minutes. The time for severe cutting and editing will have arrived. This task can be demanding and discomforting because you have conceived, developed, tested, and most likely become attached to your project and its results. Now you must wield your computer's "delete" key to select and reject among your intellectual offspring. You can ease the process a bit by reminding yourself that you really aren't sacrificing among the fittest of your brood. You merely are deciding which can go to the conference and which must stay home. Perhaps the others can go the next time. (Note that all of your publishable data can go to the conference if you take copies of a written manuscript to share with other interested professionals.)

A final note on preparing your talk: A common formula for designing and delivering a speech is to "Tell 'em what you're gonna tell 'em. Tell 'em. Tell 'em what you told 'em." Here's another tip: Do it but don't make it obvious. You'll turn 'em off.

So far, we've covered a general approach for preparing a professional talk; now let's consider more specific suggestions and criteria for each component.

Title

Keep it short, yet informative. Avoid unnecessary jargon and strings of prepositional phrases. Make a title slide that includes the title of your talk, your name, and the names of any coauthors. If appropriate, make an additional slide to acknowledge funding organizations and other significant cooperators.

Opening

You need to capture audience attention at the beginning. Remember that the room probably has several distractions, especially at the start of your talk. Begin speaking only after you have reached the podium, been introduced, properly fitted the microphone, faced your audience, and taken a deep breath.

Your words should be straightforward, clear, and forceful. They should frame your talk and aim at audience interests. Your approach should be designed to arouse those interests, create a mood, and prepare the audience for the objectives that will follow.

Here are some possible techniques:

- Place your work in the context of the overall symposium, perhaps by referring to the conference purpose or setting, to other topics or speakers on the program, or to significant work that preceded yours.

- Refer to your audience. Warm them with a subtle compliment to their interests and expertise. Don't overdo it—that can be perceived as flattery; or worse, fawning; or still worse, patronizing.

- Relate a pertinent anecdote or personal experience, or give a relevant quotation.

- Outline the background or overall problem to which your project relates. Connect it to the audience.

- Offer a comparative or contrasting analogy that leads to your subject.

- Ask a rhetorical question, then answer it. Don't ask for audience participation.

- Make a statement that they will find slightly disturbing. (Caution, choose wisely. This approach can be poorly received and backfire on you.)

Although your approach will vary with subject matter, audience, and purpose, it always should fit the occasion and your personality. Don't try to say or do something that feels forced or unnatural.

Here are some more "don'ts":

Don't

- begin speaking until you are in position and ready.

- wait for the entire audience to be ready before you begin speaking—if you do, you'll never start.

- begin by saying something inane or stupid like "I don't know how I'm going to fit all this into 15 minutes."

- mumble anything to yourself.

- get cute, clever, or hyperbolic.

- tell an off-color, tired, or irrelevant joke (but subtle, fitting humor is good).

- be late.

Here's a "do":

- Practice, practice, practice your opening, and test it on some friends. You know the aphorism: "You get only one chance to make a good first impression."

Objective

Tell the audience the specific intent(s) and purpose(s) of your work, not of your talk. They already know that the objective of your talk is to tell them about your work. From this point on, focus on your message, not on yourself. The objective(s) can be part of your opening. Create a slide that outlines each objective.

Content

What is your main message? Make it clear, well organized, and strongly developed.

- Outline your research methods on a slide or two.

- Show a couple of pictures of field sites, equipment, or whatever's relevant. Include people in the photos.

- Simplify your results, and display data in graphical form (e.g., histograms, pie charts, scatter diagrams). The more visual, the better.

- Include error bars and statistical significance whenever appropriate.

- Avoid using tables to present data. (Although data tables might be necessary and effective in your thesis or professional paper, they often are difficult for an audience that has never seen them before and has less than a minute to view and process the information. If you *must* use tables, keep them simple.)

- Include mathematical formulas, chemical reactions, or computer models and diagrams as needed, but keep the images large and as simple as possible. For complex formulas and models, consider separating them into simpler component slides, then merging them in another slide.

Transitions

Plan how you will shift smoothly from one topic to the next. A good way to accomplish this is through well chosen words in slide headings.

Visual Aids

Set up your talk in a computer program designed for creating public presentations (e.g., Microsoft's Power Point). These programs are loaded with specialized features for designing and producing professional-grade slide shows. *Be aware that the program and operating system you use must be compatible with the venue at which you will present. Check with the sponsoring organization for specific requirements and limitations.*

To be effective, each slide in your show must meet the criteria expressed by three words: *relevance, refinement, readability.*

Relevance: Each slide should have a clearly defined purpose and should logically, precisely, and succinctly contribute to the substance of your talk. If a slide's content is only peripherally important, delete it.

Refinement: Design each slide to present your message with purity, polish, and precision. Use caution. Don't get carried away by superfluous ornamentation, dazzling color, and zooming action. Trying to add variety and attract attention though glitz and gingerbread might sell soap, but it won't impress a roomful of hard-core thinkers, no matter how good natured they might be. It will only distract and annoy them. Beside, they all own the same computer program; they will have seen it all. (An exception can be made for talks *intended* to demonstrate a new computer program.)

Readability: If viewers can't read it, it's no good. Readability requires large fonts, strong contrast between print or graphics and background, simplicity, and no clutter. Readers should be able to focus immediately on key words. Don't force them to read sentences to grasp your message.

Here are a few more specific pointers:

- Use an unadorned font, one that projects sharply and is immediately recognizable. Avoid unusual, decorative, or obtrusive styles as well as serif lettering. In most cases, if it's not suitable for a traffic sign, don't use it.

- Choose color schemes that will make your message stand out under any lighting conditions. Ensure high contrast between the foreground and background, and shun pastel colors except perhaps for backgrounds. Avoid gaudy color schemes and colors that clash. If you don't have a sense for color coordination, ask someone who does.

- Keep slide designs simple and balanced. Ask yourself, What is the one main message I want to convey with this slide? Then design the slide accordingly. Although slides need not be austere, they should be free of irrelevant distractions. Most importantly, do not imitate TV programs that display multiple images and running banners along with two or three talking heads. Although a production of this sort might briefly capture a fickle audience, it's more likely to irritate sophisticated professionals. (Note that this recommendation does not rule out the use of amusing or otherwise amiable images. These can set your audience at ease. Just be sure they are relevant and not overdone.)

- Use animations judiciously. As with image designs, animations can either enhance your talk or distract and annoy your audience. Animations should not draw attention to themselves; rather, they should help your audience follow your main points.

Closing

Briefly highlight your key points, namely your major findings and conclusions. Carefully consider the main message you want your audience to leave with. If you can't write this message plainly on the back of a business card, you need to give it more thought.

Preparing Yourself for the Talk

Physical Manner

Posture

Weeks before your presentation, begin practicing your posture, whether that be standing up straight or sitting as erect as possible. Look at yourself in a full-length mirror and think about your body one part at a time. Concentrate first on your feet and legs, then move up through hips and midsection to chest, shoulders, and neck. If you are fully able-bodied, practice standing comfortably and shifting your weight slightly from one leg to the other without locking your knees or slumping your hips. (If you allow one hip to slump and one knee to lock, chances are your shoulders will sag slightly and so will your voice.) Pull your chest up and shoulders back. The point is to present yourself with confidence, no matter what your body style or condition. If you don't exercise regularly, now's the time to start.

Movements and gestures

Practice your "body language." If you move your feet, pick them up; do not shuffle. Don't stuff your hands in your pockets or fold your arms. Watch a couple of TV lawyers or practiced politicians to get ideas of what to do with your hands—just don't try to copy their exaggerations too literally. The object is to use controlled movements and gestures to help emphasize verbal points, but in a way that the audience doesn't consciously notice. Keep in mind that your freedom of movement might be limited at a podium.

Eye contact

The more eye contact you can maintain with your audience, the better. Practice with yourself before a mirror, and think about it occasionally during conversations.

Voice

In addition to preparing your mind and body, you should begin preparing your voice. You easily can practice projection, speed, flexibility, and clarity every day during normal conversation.

Rehearse your professional talk by speaking to a recorder set up across a room. Pay attention to your volume, but also listen carefully to your natural tempo, intonations, and pronunciations. Speak as loudly as you can and still be comfortable. (But when the real time comes, be careful not to yell into a microphone.) Try varying the tempo slightly to help emphasize important points. If you hear yourself speaking in a monotone, add emphasis to key words. (Listen carefully to the speaking styles of several professional newscasters. You won't be able to copy them—and you probably shouldn't—but you'll get some ideas.) Finally, listen to your pronunciation. Do you tend to slur or clip the endings of certain words? If so, correct the problem. Ask a trusted colleague to sit through one of your rehearsals and offer criticism.

Language and Speech

Choose your words thoughtfully. Use correct grammar, and speak succinctly. If you're unsure about definitions or grammatical rules, look them up. Use technical terms whenever appropriate but avoid unnecessary jargon. (You are speaking to a professional audience—they know the lingo but won't be impressed by pretentious language.) Practice correct pronunciation of everyday words as well as technical terms (e.g., the word is "to" not "tuh"). Watch out not to slur or clip words (e.g., "gonna" for "going to," "talkin'" for "talking"). Never use overly informal phrases or slang unless the words are necessary and germane to your subject. Avoid indefinite references (e.g., the generic "they" when you mean "researchers"). Never refer to your audience as "you guys" or other casual, irreverent denominations.

Miscellaneous Suggestions

As you create your visual aids and store them in your computer *back them up!* Print full-size, color copies on regular paper and another set of smaller copies on note cards. The note cards are handy for practicing your talk without a computer. For example, you can refer to them during walk-and-talk practice sessions.

What will you do if you suffer a technical malfunction (e.g., the computer freezes) while on stage? Hmm. Good thing you practiced and have those back-ups in your brief case and handy note cards in your pocket.

Here are a few more things you'll have to do on the day of your talk:

- Most professional meetings have a practice room—use it.
- Check out your presentation room well before your session starts.
- Find your moderator, introduce yourself, and make sure she or he knows how to introduce you properly.
- You probably will not use your own computer. Learn to download your talk from your storage disk or internet connection to the presentation computer.
- Know how to find and open your talk in the presentation computer's files.
- Practice starting, advancing, and rerunning your slides.
- Locate the pointer and practice using it.

Finally, the absolute best ways to gain confidence and improve your abilities in public speaking are practice and experience, followed by more practice and experience. Other graduate students in your program will be going through the same agonies and ecstasies. Get together with them and your major professors for practice and critique sessions.

Delivering the Talk

If you've followed all the advice so far, you're ready to go. The better prepared you are, the more relaxed and confident you'll feel and the better you'll do.

Here are a few reminders and last minute pointers:

- Dress for success. You will feel more in control and command greater audience attention and respect if you do. As you gain experience, and if eventually you become a "member of the club," you might dress less formally. Even then, a bit more formal than the audience is a good rule.

- Speak to your audience, not to the projection screen or computer. Do your best to maintain eye contact with members of the audience. Speak around the room to small groups and individuals as though they were friends or colleagues. They will be more attentive and sympathetic. Remember that the audience is not adversarial. They are on your side and want you to do well. They're interested in what you have to say. If they weren't, they wouldn't be there.

- Voice projection to an audience is not simply a matter of volume, it also depends on direction and clarity. So speak up, speak directly to your audience, and pronounce your words crisply. Keep your voice up through the end of each sentence.

- When you finish with a slide, pause for a second or two before advancing to the next frame. Audience members will be thinking about the image and about what you've just said. Give them a chance to synchronize their senses.

- Do not leave a slide on the screen after you have finished with it. A displayed image should go with your words.

- If "ums" start to come out of your mouth, close your mouth.

- Maintain your posture. If your body sags, your message likely will sag with it.

- Don't hand out anything at the start of your talk. You may make reprints or handouts available when you're finished.

- Watch your time. If you should happen to run short, cut something so that you can finish on time. Do it discreetly; don't make a show of what you're leaving out.

The major preparation and delivery points presented above are listed in the Oral-Report Evaluation Form (Fig. 6-1). Feel free to photocopy the form, share it with colleagues, and use it to critique each other during practice sessions. The form might have more detail than you want; still, it's a good check list for your practice audience and for you as a speaker. Be honest in your evaluations, and try to express criticisms and suggestions in a positive manner.

Do well and feel good about it.

Self Analysis

After you've given your presentation, in practice or for real, prepare a one-page self-analysis. Do it as soon as you can after your talk, and write it into your personal diary or free writing notebook. Consider the following points:

Oral-Report Evaluation Form

Speaker _____   Reviewer _____

1. Title _____
2. Physical appearance _____
3. Opening _____
4. Objective _____
5. Physical manner
   A. Posture _____
   B. Movement and gestures _____
   C. Eye contact _____
   D. Composure _____
6. Voice
   A. Projection and clarity _____
   B. Flexibility _____
   C. Speed _____
7. Language and speech
   A. Grammar _____
   B. Pronunciation _____
   C. Appropriateness and style _____
8. Content
   A. Organization _____
   B. Strength and development _____
   C. Clarity _____
9. Transitions _____
10. Visual aids
    A. Readability _____
    B. Layout _____
    C. Color scheme and contrast _____
    C. Relevance _____
    D. Animations _____
    D. Effectiveness _____
11. Closing _____
12. Use of time _____

What were the best parts of the presentation?

Suggestions for improvement?

Figure 6-1. Oral-report evaluation form.

- Begin with what you did well. What parts of your presentation do you think were most effective? With which ones are you most satisfied?

- What could use improvement? Consider the criteria presented earlier in this chapter and in the Oral-Report Evaluation Form, as well as any other points you might deem relevant.

- What are your goals for improving your next presentation? Be specific in your response.

- What do you think you can do to overcome difficulties and improve your comfort level as well as your message and delivery?

Poster Presentations

At professional conferences, scientists often present their work on printed posters rather than orally. Posters, which typically measure about 1.5 meters wide by 1 meter tall, explain the research through illustrations and text.

A poster can contain about the same information as a 15-minute talk, but the information is presented in far fewer words. The most important points should be made graphically, otherwise most people, unless they are particularly dedicated to the topic, won't pay attention. Even enthusiastic scientists can become saturated and lose interest when surrounded by verbiage-crammed panels in a room the size of a warehouse. They'll look for the posters that grab their attention and pique their interest and then talk with friends while they ignore the rest. Nonetheless a good poster can generate discussion that goes on far longer than in any structured talk.

Planning a talk and planning a poster require about the same effort and many of the same activities. You must focus the content, cut the extraneous material, and target your audience. If you can't write your main message on the back of a business card, you likely will have trouble composing a poster.

Numerous internet sites offer excellent (and some not-so-excellent) sugges- tions for poster presentations. Simply type "poster presentations" or "research posters" into a good search engine and you'll find plenty of instruction. Details are not given here because sites and computer software for making posters change frequently.

Figure 6-2 outlines some criteria for a research poster. (Fig. 6-2 would not be a good poster by itself because it lacks graphical illustrations and a focal point.) Here are a few additional pointers:

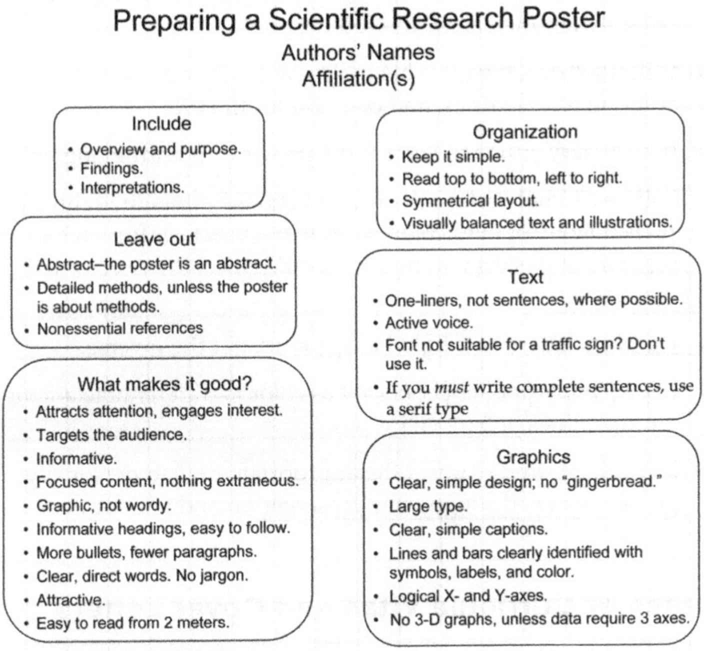

Preparing a Scientific Research Poster
Authors' Names
Affiliation(s)

Include
- Overview and purpose.
- Findings.
- Interpretations.

Leave out
- Abstract—the poster is an abstract.
- Detailed methods, unless the poster is about methods.
- Nonessential references

What makes it good?
- Attracts attention, engages interest.
- Targets the audience.
- Informative.
- Focused content, nothing extraneous.
- Graphic, not wordy.
- Informative headings, easy to follow.
- More bullets, fewer paragraphs.
- Clear, direct words. No jargon.
- Attractive.
- Easy to read from 2 meters.

Organization
- Keep it simple.
- Read top to bottom, left to right.
- Symmetrical layout.
- Visually balanced text and illustrations.

Text
- One-liners, not sentences, where possible.
- Active voice.
- Font not suitable for a traffic sign? Don't use it.
- If you *must* write complete sentences, use a serif type

Graphics
- Clear, simple design; no "gingerbread."
- Large type.
- Clear, simple captions.
- Lines and bars clearly identified with symbols, labels, and color.
- Logical X- and Y-axes.
- No 3-D graphs, unless data require 3 axes.

Figure 6-2. Preparing a scientific research poster.

- Allow at least a month to prepare, more if you have coauthors. (Of course, you'll have other things to do during that month.)

- Before you begin, find out about the venue, amount of space available, and mounting restrictions.

- Check printing costs for the poster and for small reprints to hand out. Did you include these costs in your research budget?

- Try to give your poster a center of interest—a focal point that captures viewers' attention, encourages them to read other parts, and coaxes them back to the main point.

- Don't bother putting an abstract on your poster—your poster *is* an abstract. (Note that conference rules might require that a standard abstract be submitted separately from the poster.)

- Display conclusions prominently. Many readers will look for these first.

- Favor informative headings over generic ones. For example, if your results showed a population increase, a heading could read "Population Increases" instead of simply "Results."

- With font size, bigger is better. Use at least 24-point for text, 36-point for headings, and 60-point for the title.

- If necessary, guide viewers with arrows or pointing fingers.

- Follow a simple color scheme, and choose colors thoughtfully. A cacophony of loud colors will distract and confuse, while a plethora of pastels will not stand out, especially to color-blind viewers.

- Use dark letters on a light background, not the inverse.

- Use symbols and labels as well as color to differentiate among multiple lines or bars on graphs.

- Be available and ready at the appropriate time to personally guide viewers through your presentation and answer questions.

Résumés or Curricula Vitae and Cover Letters

You can find good information about applying for jobs, writing résumés, and interviewing in a multitude of books, articles, and websites. Your university's placement office or student services center probably can give you a stack of helpful pamphlets. But since you're here and this chapter is about presenting yourself, here are some suggestions about preparing a résumé and writing a cover letter.

The Résumé or Curriculum Vitae

A résumé or curriculum vitae outlines your background and qualifications for a professional position or further academic study. Its format and content can be made to fit the purpose, as long as the information remains accurate and doesn't mislead. For example, you may emphasize education over experience or vice versa; you could include certain hobbies in one version but not in another; and you might include or omit certain personal information. When considering personal information, ask yourself, Is my age, religious affiliation, or political party any more relevant than my skin color? It might or might not be, but consider the matter prudently. Above all, ensure that your résumé is well formatted and free of errors.

As you gain experience, make several versions of your résumé to fit different purposes. Include in one version all the details you ever will need. This is your master copy. You probably won't send it out but will use it to make adaptations. The final editions should run about two to three pages, depending on their purpose and your experience. Sometimes prospective employers, especially those who must flip through hundreds of applications, require single-page résumés. If so, follow their instructions, but in most cases don't sell yourself short by unnecessarily minimizing your qualifications. By mid-career, your résumé could run a half dozen pages, especially if you have publications.

Any good bookstore will carry several "how-to" books on résumé preparation. And of course you'll find more than enough information on the internet. Check them out, bearing in mind that you'll find bad advice along with the good. Meanwhile here are a few things you should know about preparing and handling your résumé:

- Place your contact information at the top. Make it easy to find and easy to read.

- Make the organization immediately clear; use bold headings to separate major categories.

- Write in outline form, not full sentences.

- Use a simple, easy-to-read font.

- For more than one page, print them single sided.

- Unless instructed differently, send only one copy to each recipient.

- Do not attach a photograph of yourself.

- Follow the instructions in the position announcement. Check the organization's website for additional information.

- Large companies and government agencies have a human resources (HR) department staffed with people who know the protocols and hiring rules. After doing your homework, contact someone there if you need more information about procedures and addressing your cover letter.

- If the organization has an application form, fill it out completely. Don't simply fill in a blank with "See attached."

- Try to submit your whole application package at once, rather than in separate mailings.

Here's what to include in a résumé:

At the Top
- Your name
- Address
- Telephone numbers
- E-mail address
- Citizenship or residency status

If you have a temporary address and phone number as well as permanent ones, make that clear and indicate the dates during which the temporary information applies.

Education
Start with the most recent or advanced degree. List each degree title, major, minors, year of graduation, and university. If you are a student, give your status and anticipated graduation date. Include thesis and project titles. You might or might not want to list institutions you attended without earning a diploma, but include them in your long version.

Experience
Begin with the most recent and work back. Include dates (years might be close enough) of employment, position title and rank, employer, and employer's address. A brief (one or two lines) job description or list of duties often is helpful. Focus on your accomplishments, but don't appear boastful. Leave out unrelated and trivial experiences, unless they illustrate your willingness to assume responsibility and take initiative. Don't leave conspicuous unexplained gaps that could raise unnecessary questions.

Honors, Awards, Licenses, and Certificates
If you received the Medal of Freedom, put it in. If you were voted cutest baby at the county fair, leave it out. Dean's list and scholarship awards should go in, as should professional certifications and other certified skills. Don't overlook your driver's license if you have one.

Languages
List any supplementary languages you might know, and rate your proficiency in speaking, reading, and writing.

Professional Organizations
List all professional organizations to which you belong. If you hold an office or are particularly active, include that information. Don't forget that some student clubs are affiliated with national professional societies.

Publications and Presentations

List any that might help you get the job.

Professional Goals and Interests

A good goal statement always strengthens a résumé. Make it enthusiastic but not exaggerated. Show self-confidence, not self-importance. Be personable, not portentous. Above all, make it sincere.

Hobbies and Other Personal Interests

List any that might help you get the job. Don't forget foreign travel if that is in any way relevant.

Cover Letters

Although a strong cover letter probably won't land you a job, it can mean the difference between being interviewed and being passed over. If your résumé or job application shows that you meet minimum requirements, someone will read your cover letter.

Here are some tips for the letter:

- Before you write, learn about the organization and position for which you are applying.

- Follow instructions in the position announcement.

- Follow a standard business-letter format.

- If possible, address it to a person. If the position announcement doesn't tell how to address the letter, contact the Human Resources department to find out. If that fails, address it to a position title, but not to "Dear Sir."

- Unless you want to be a camp counselor, don't begin with "Hi, my name is..." Your name and signature should be properly placed at the bottom of your letter.

- Specify the position for which you are applying.

- Tailor your letter to the organization and position for which you are applying. Don't simply write boilerplate—stock sentences that say nothing informative and that are no different from what every other applicant could write.

- Put some life in it—be enthusiastic but not effusive.

- Don't bother applying for something you really don't want.

CHAPTER 7

When It's All Over

What's This Chapter About?

Just what it says

Is that all there is?

Peggy Lee, 1960's pop tune based on Thomas Mann's story *Disillusionment*

When you've completed the course work, submitted the final thesis, passed the oral defense, and hung the degree on your wall, you'll think it's all over. But probably it won't be because the experience will linger. Among you, some will march on to the next phase of life; others will trudge. Some will party all night, buoyed by their spirits; others will stand vigil all night, guarding their spirit.

As a few of you look immediately forward, many more will look back, reflecting on how they could have done it differently. Some measure of dissatisfaction with their performance or their final product might elicit a tinge of regret for tasks left undone or opportunities not taken. A range of alternative approaches and additional analyses might run through their minds—routes and methods which by now are irretrievable and fruitless to ponder, except for their value as learning experiences.

If this happens to you (and likely it will), bear in mind that learning is what it's all about. The process you went through—learning to master a discipline, think scientifically, and solve problems—has afforded you intellectual strength and competence beyond the norm. So do your best to leave the trail of "should haves" behind and strike for the path ahead. Your knowledge of scientific process has become your stock-in-trade, and it will serve you well in nearly all aspects of life. It is, after all, the true product of your efforts.

Another type of response will creep over a sizable proportion of you. It overlaps with the "should-have syndrome" but extends deeper than mild dissatisfaction and rechewing experiences. Those of you who suffer this will look back exhausted and ask, Was it worth it? As the day you've looked for, labored for, and sacrificed for comes and goes, you will sense an uncertainty—an uneasiness that's not quite definable. You'll be as bewildered as a moose calf who looks up from browsing willow shoots one late summer day to find his mother gone. Moments before, she casually had drifted over a knoll, leaving him forever on his own. But unlike the moose, you probably won't know exactly what the problem is. The reasons are less obvious.

OK, enough of the moose, we don't want to overdo that little allegory! So now as your author, I'll violate one of Strunk and White's rules and come out from behind the keyboard. Over the years, I've detected that a feeling of post-graduation letdown is fairly common among those of us who have intensely pursued advanced degrees. What's wrong? We should revel in relief, joy, and satisfaction. But too often we don't. Too often we descend into an anticlimactic void, a sort of emptiness in the soul.

These feelings appear to be more intense with completion of the Ph.D. degree than with the master's degree. If this is true, is it solely because the Ph.D. is more intense, more mentally exhausting? Or is it also because the Ph.D. is the terminal degree, the end point? Where can we go from here? What else can we prove about our intellectual prowess? Where will our guidance come from? Are we really on our own?

I'm not sure of the reasons for this sometimes disappointing descent, but I have a few ideas—hypotheses if you will. So here they are. I hope you find them helpful and reassuring.

Before completing graduate school, most of us live our lives with defined goals, which usually are set by someone else or by circumstances. These goals commonly take a set number of years to attain: we graduate from high school, earn a bachelor's degree, perhaps serve a hitch in the military or public service, or take a job that we know is temporary before going on to graduate school. Then perhaps the master's degree is another two-year program, and the Ph.D. another four. Each of these has a time line; the end stays reasonably in sight. These are goals for getting on with life.

When the pursuit of the terminal degree ends, the short-term goals end with it. For some graduates, life beckons like a flag of liberty and opportunity; for others, it looms as an expanse of open ocean to a rowboat; and to still others, it confuses like a maze of magnets to a compass. And no matter what our situation, without quite understanding, we're just a bit overwhelmed.

Even if we land the perfect job and start building a career, we might wonder if we should have done something differently. Should we have taken a different path—one less tortuous, one having a higher benefit-to-cost ratio? Maybe we think about student loans that are due, or the family time we've missed, or how we could have been seeking our fortune more productively. Instead, we enslaved ourselves in academia. We strained our resolution as well as our intellect.

So the realization and full satisfaction of your accomplishment probably will not come instantly. It likely will seep in over the years as you gradually reap and recognize the benefits. But what should you do in the meantime?

If you feel this sort of letdown, one good solution is to learn from experience and set new goals. Picture something definite you'd like to accomplish in the short term, and outline a three-year plan for getting there. It doesn't have to be grand, just doable. The idea will give you focus, and with time the uncertainty and distress will dissipate. As the trials become more distant, you'll be able to look back, understanding and accepting that the path you took was worth it. Continue, and it will lead to new opportunities for variety, intellectual stimulation, and fulfillment.

If you choose a career in scientific research, your success will not depend on how well you follow methodological rules (Feyerabend, 1975). Rather, it will depend on your desire to know how things work, your creativity, and your willingness to go beyond where you've been before.

Even if you do not pursue research, your training will help you appreciate the elegance of good science. You will understand that in education, one thing more disquieting than an unanswered question is an unquestioned answer. You will realize that to accept without challenge or to reject without consideration is negligence. And you will come to know that studies of scientific philosophy and process, along with a few other intense human endeavors, help us to think, to experience, and to learn while they fuel our appreciation and passion for life and for intellectual freedom.

And please, when you do graduate, take your knowledge, thinking skills, and problem-solving abilities with you. Then put them to good use. The world needs them.

REFERENCES

American Society of Agronomy. 2011. Publications handbook and style manual. ASA, CSSA, and SSSA. Madison, WI. (Available at https://www.crops.org/publications/style)

Averill, J.R., and E.P. Nunley. 1992. Voyages of the heart: Living an emotionally creative life. The Free Press, New York.

Bunch, B., and A. Hellermans. 2004. The history of science and technology. Houghton Mifflin Co., Boston.

California Code of Regulations. n.d. 5 C.C.R. title 5, div. 5, ch. 1, subch. 2, art. 7, sec. 50510, para. 3.

Carnap, R. 1962. The aim of inductive logic. p. 303–318. In E. Nagel, P. Suppes, and A. Tarski (ed.) Logic, methodology and philosophy of science. Proc. Int. Cong. Logic, Methodology, and Philosophy of Science. 1960. Stanford Univ. Press, Stanford, CA.

Coghill, A.M., and L.R. Garson (ed.) 2006. The ACS style guide: Effective communication of scientific information. 3rd. ed. Oxford Univ. Press, Oxford.

Cohen, J., and G. Medley. 2005. Stop working and start thinking. Taylor and Francis, New York.

Cottingham, J. 1988. The rationalists. A history of Western philosophy, vol. 4. Oxford University Press. Oxford.

Dodd, J.S. (ed.) 1997. The ACS style guide: A manual for authors and editors. American Chemical Society. Washington, DC.

Evans, J., and A. Feeney. 2004. The role of prior belief in reasoning. In J.P. Leighton and R.J. Sternberg (ed.) The nature of reasoning. Cambridge University Press, Cambridge.

Feyerabend, P. 1975. Against method. New Left Books, London.

Feynman, R.P. 1999. The pleasure of finding things out. Perseus Books. Cambridge, MA.

Fleck, L. 1979. Genesis and development of a scientific fact. University of Chicago Press, Chicago.

Gauch, H.G., Jr. 2003. Scientific method in practice. Cambridge Univ. Press, Cambridge.

Gladon, R.J., W.R. Graves, and J.M. Kelly. 2011. Getting published in the life sciences. Wiley-Blackwell, Hoboken, New Jersey.

Gould, S.J. 1977. The validation of continental drift. p. 160–167 In Ever since Darwin: Reflections in natural history. W.W. Norton, New York.

Gould, Stephen Jay. 1983. Evolution as fact and theory. p. 253–262 In Hen's teeth and horse's toes. W.W. Norton, New York.

Hawking, S.W. 1988. A brief history of time. Bantam Books, New York.

Hempel, C.G. 1966. Philosophy of natural science. Prentice Hall, Upper Saddle River, NJ.

Hume, D. [1739] 1967. A treatise of human nature. Book 1. Of the understanding. Clarendon Press, Oxford.

Hurley, P.M. 1968. The confirmation of continental drift. Scientific American 218 (4):52–64.

Huxley, T.H. 1902. Science and education: Essays. D. Appleton, New York.

Huxley, T.H. [1894] 2004. Collected essays, vol. viii: Discourses, biological and geological. Elibron Classics.

Huxley, T.H.. [1900] 2006. Huxley's autobiography and essays. Cosimo Classics, New York.

Johnson-Laird, P.N. 2006. How we reason. Oxford University Press, Oxford.

Kanare, H.M. 1985. Writing the laboratory notebook. American Chemical Society, Washington, D.C.

Kant, I. [1787] 1900. The critique of pure reason. Wiley Book Co., New York.

Kotarbinska, J. 1962. The controversy: Deductivism versus inductivism. p. 265–274. *In* E. Nagel, P. Suppes, and A. Tarski (ed.) Logic, methodology and philosophy of science. Proc. Int. Cong. Logic, Methodology, and Philosophy of Science. 1960. Stanford Univ. Press, Stanford, CA.

Kuhn, D., E. Amsel, and M. O'Loughlin. 1988. The development of scientific thinking skills. Academic Press, San Diego.

Kuhn, T.S. 1996. The structure of scientific revolutions. 3rd ed. University of Chicago Press, Chicago.

Kuyper, B.J. 1991. Bringing up scientists in the art of critiquing research. BioScience 41:248–250.

Ladd, G.W. 1987. Imagination in research: An economist's view. Iowa State University Press, Ames.

Lakatos, I. 1978. The methodology of scientific research programmes. Cambridge University Press, Cambridge.

Leedy, P.D. 1974. Practical research. Macmillan, New York.

Leedy, P.D., and J.E. Ormrod. 2005. Practical research: Planning and design. Pearson Prentice Hall, Upper Saddle River, NJ.

Medawar, P. B. 1967. The art of the soluble. Methuen, London.

Medawar, P.B.. 1969. Induction and intuition in scientific thought. American Philosophical Society, Philadelphia.

Mill, J.S. [1843] 1874. A system of logic ratiocinative and inductive. Longmans, Green, London.

Nelson, M.P., and J.A. Vucetich. 2009. On advocacy by environmental scientists: What, whether, why, and how? Conservation Biology 23(5):1090–1101. doi: 10.1111/j.1523-1739.2009.01250.x

Oaksford, M., and N. Chater. 2007. Bayesian rationality: The probabilistic approach to human reasoning. Oxford University Press, Oxford.

Pace, N.R. 2006. Time for a change. Nature 441:289.

Pears, D. 1990. Hume's system: An examination of the first book of his Treatise. Oxford Univ. Press. New York.

Peters, E. 1988. Inquisition. Univ. Calif. Press, Berkeley.

Plummer, C.C., D.H. Carlson, and D. McGeary. 2007. Physical geology. McGraw-Hill, Boston.

Popper, K.R. 1959. The logic of scientific discovery. Basic Books, New York.

Popper, K.R. 1962. Conjectures and refutations: The growth of scientific knowledge. Basic Books, New York.

Russell, B. 1945. A history of Western philosophy. Simon and Schuster, New York.

Salmon, W.C. 1988. Rational prediction. p. 433–444. *In* M. Curd and J.A. Cover (ed.) 1998. Philosophy of science: The central issues. W.W. Norton, New York.

Seuss, Dr. [T.S. Geisel]. 1990. Oh, the places you'll go. Random House, New York.

Stenning, K., and P. Monaghan. 2004. Strategies and knowledge representation. p. 129–168. *In* J.P. Leighton and R.J. Sternberg (ed.) The nature of reasoning. Cambridge University Press, Cambridge.

Strunk, W., Jr., and E.B. White. 2000. The elements of style. 4th ed. Macmillan, New York.

Taskey, R.D., C.L. Curtis, and J. Stone. 1989. Wildfire, ryegrass seeding, and watershed rehabilitation. p. 115–124. *In* N.H. Berg (tech. coord.) Proc. Symp. Fire and Watershed Manage. Sacramento, CA. 26–28 Oct. 1988. USDA Forest Serv. PSW For. Ran. Exp. Sta. Gen. Tech. Rep. PSW–109.

Tweney, R.D., M.E. Doherty, and C.R. Mynatt (ed.). 1981. On scientific thinking. Columbia University Press, New York.

University of Chicago Press. 2010. The Chicago manual of style. 16th ed. University of Chicago Press, Chicago.

Valiela, Ivan. 2009. Doing science: Design, analysis, and communication of scientific research. Oxford University Press.

Whewell, W. [1847] 1967. The philosophy of the inductive sciences, founded upon their history. Parts 1 and 2. Frank Cass and Co. London.

Whewell, W. [1860] 1971. On the philosophy of discovery. Lenox Hill, New York.

White, A.D. 1898. A history of the warfare of science with theology in Christendom. Vol. 1. D. Appleton, New York.

Winn, L.J. 1997. Making effective oral presentations. p. 367–392. *In* J.S. Dodd (ed.) The ACS style guide: Manual for authors and editors. 2nd ed. American Chemical Society. Washington DC.

Woese, C.R. 2004. A new biology for a new century. Microbiol. Mol. Biol. Rev. 68(2):173–186.

Woese, C.R., O. Kandler, and M.L. Wheelis. 1990. Towards a natural system of organisms: Proposal for the domains Archaea, Bacteria, and Eucarya. Proc. Natl. Acad. Sci. USA 87:4576–4579.

Woodhouse, M.B. 2006. A preface to philosophy. 8th ed. Wadsworth, Belmont, CA.

Woolhouse, R.S. 1988. The empiricists. A history of Western philosophy. Vol. 5. Oxford University Press. Oxford.

Zalta, E.N. (ed.) 2004. Stanford encyclopedia of philosophy. http://plato.stanford.edu/archives/fall2004 (verified 23 Aug. 2011)

Zilsel, E. 1941. Problems of empiricism. p. 803–844. *In* O. Neurath, R. Carnap, and C. Morris (ed.) Vol. 2 of Foundations of the unity of science: Toward an international encyclopedia of unified science. University of Chicago Press, Chicago.

INDEX

Finally,

Don't forget to swing hard in case you hit the ball.

Woodie Held

Printed and bound by CPI Group (UK) Ltd, Croydon, CR0 4YY

27/10/2024

14580334-0001